STO

12-6-78

BIOLOGY: ITS HISTORICAL DEVELOPMENT

BIOLOGY: ITS HISTORICAL DEVELOPMENT

by
Howard B. Baumel

Philosophical Library
New York

Copyright, 1978, by Philosophical Library, Inc.,
15 East 40 Street, New York, N.Y. 10016
All rights reserved
Library of Congress Catalog Card No. 77-087937
SBN 8022-2217-X
Manufactured in the United States of America

In memory of Morris Baumel and for Grace, Jana and Jennifer.

CONTENTS

THE BEGINNING OF BIOLOGY   1

Earliest Influences
Greek, Egyptian and Roman Contributions
The Arabic Role
Renaissance Science

THE RISE OF MODERN BIOLOGY   9

Circulation
Microscopes and Cells
Photosynthesis
Physiology
Nutrition
Hormones

THE CONQUEST OF DISEASE   29

Vaccination to Antibiotics
Virology
The Search for a Polio Vaccine

THE ORIGIN OF LIFE   49

Disproving Spontaneous Generation
Returning to Spontaneous Generation

THE DEVELOPMENT OF GENETICS 55

Pre-Mendelian Inheritance
Mendel
Rediscovery
Garrod—Another Ignored Contribution
The Rise of Modern Genetics
The Genetics of Sickle Cell Anemia
Discovery of the Hereditary Material
Genetics in the Soviet Union
Synthesizing Nucleic Acids
Recombinant DNA and Genetic Engineering

DEVELOPING A THEORY OF EVOLUTION 81

Setting the Stage for Darwin
Charles Darwin
The Social and Intellectual Climate
Supporting Darwin: Huxley, Wallace and Gray
Darwin's Later Work
Darwinism Today

THE BEGINNING OF BIOLOGY

Earliest Influences
Greek, Egyptian and Roman Contributions
The Arabic Role
Renaissance Science

Earliest Influences

Preserved clay models of organs of the body from Babylon, the earliest home of human civilization, indicate that the knowledge of anatomy there was considerable. To the Babylonians, the heart was regarded as the organ of intelligence and the liver as the organ of circulation.

In Egypt, the other oldest civilized country of the west, the art of healing was based on actual observation rather than superstition. The Egyptian religious practice of preserving dead bodies offered opportunities for the acquisition of anatomical knowledge.

The Hebrew conception of nature is preserved in the Old Testament and in a supplement to the Bible called the Talmud. Since these works contain many laws regulating personal cleanliness and hygiene, they could actually be regarded as the first public health documents. For example, swine were considered unclean and there were explicit instructions for handling lepers and admonitions against living in cities without bathhouses and physicians. The Bible also includes a description of every known group of animals as well as observations about their behavior.

Hindu science emphasized mathematics and Chinese lore involved two opposing vital forces, Yang and Yin, which were used to explain the functions of the organs of the body. The Chinese also ascribed medicinal value to many plants.

Greek, Egyptian and Roman Influences

Scientific knowledge, as we recognize it, is considered to have begun with the ancient Greeks. A medical school arose on one of the Greek islands as early as 600 B.C. The greatest member of this school was Hippocrates (460-377 B.C.) who is regarded as the father of medicine. He dissected dead animal bodies and believed that the human body was composed of four basic elements - fire, water, air and earth. Four humors or juices - blood, phlegm, yellow bile and black bile corresponded to these elements.

Aristotle (384-322 B.C.) was possibly the greatest of all Greek men of science and succeeded in covering all human knowledge. He classified animals, investigated their development from the egg, differentiated between animals that reproduced sexually, asexually and by spontaneous generation and saw a "ladder of nature" with organisms progressing from lower to higher forms.

The Greek scientific period ended after the death of Alexander the Great. Then Egypt fell to Ptolemy, one of Alexander's generals, who founded a dynasty that lasted about three hundred years, ending with Cleopatra. There was much interest in science during this period and Alexandria became the center of the scientific world. A museum was established there that could be compared to a government research laboratory complex or university of today. It had botanical and zoological gardens and ten halls filled with manuscripts. The museum sheltered distinguished scholars who were engaged in translation or criticism of earlier and contemporary works and in writing new manuscripts.

After Cleopatra's death, Egypt passed under the power of Rome. Next to Aristotle, the most influential of biologists of classical antiquity was the Roman Pliny (A.D. 23-79). His famous work was *Natural History* and for 1,500 years this was the main source of man's knowledge about nature.

After Pliny, another important investigator of antiquity was the Greek physician Galen (A.D. 130-200) who practiced in Rome. He dissected animal bodies and in addition to describing the anatomy, he also attributed functions to the organs. Although Galen did not perform human dissections, his work remained the undisputed, authoritative reference on human anatomy for some 1,400 years, until the time of Vesalius. His work was held in such esteem that for centuries it was regarded as infallible and no detail which he mentioned was allowed to be altered, the result being that his errors were passed on to subsequent scientists.

Galen was working at the time that the influence of Christianity was increasing. He had developed the idea that every organ in the body was created by God in the most perfect possible form. This attitude fitted in very well with that of the Christianity of the time and of the ages which immediately followed it and helps to account for the extraordinary respect accorded Galen's works.

The Dark Ages (A.D. 200-1200) followed Galen's time and this period was characterized by a lack of biological activity that lasted for many centuries. That the Romans were an essentially practical people, devoting their best energies to the governing of their empire, is a factor that is believed to have contributed to the decline of ancient science. Science was applied to agriculture, medicine and sanitation and theoretical investigations were neither appreciated nor encouraged.

The Arabic Role

The seventh century saw the rise of Islam. By the ninth century, intellectual leadership had passed to the Arabic people and it remained with them until the thirteenth century. During this period, the Greek works were being translated into Arabic. However, many of the Greek manuscripts from the libraries of Alexandria, Athens and Rome found their way into Christian monasteries where they were translated into Latin. As a result, the Greek contributions to science were not lost during the Dark Ages. Then, in the eleventh century, the Arabic translations of the original Greek works were translated into Latin.

Renaissance Science

The rediscovery of Greek culture through these Latin translations created the desire to make use of this newly acquired Greek knowledge. One direct consequence was the founding of the university. The oldest was the one established at Bologna where human dissection began when the legal investigation of murder required the examination of the dead bodies. The study of human anatomy received additional impetus from the work of the Renaissance artists who made their own dissections in order to enable them to more accurately portray body structures.

The new medical schools of the Renaissance organized public dissections as a means of providing the medical students with the opportunity to see the body's internal organs and also as a way of making use of the few human bodies that were available for dissection. Technicians were used to do the actual dissections because most of the anatomy professors, who were trained in the scholastic tradition of the day, had never actually performed dissections and looked upon the process as being undignified. The professor usually sat at a distance from the operating table in an arena-like room and read aloud Galen's writings describing the specific organs that the dissector was working on.

With the invention of printing during the Renaissance, the ancient Greek and Latin manuscripts became much more accessible. In addition, scientists could now record their results on printed pages which would be widely distributed, enabling investigators to start new research where the previous work had stopped and to compare the experimental results of other scientists.

It remained for Vesalius (1515-1564) to correct many of Galen's errors. He is considered to be the father of modern anatomy and his book, *The Fabric of the Human Body*, is viewed as one of the first great works of modern science. Vesalius believed that knowledge could be acquired only through research and that new discoveries could be accepted only after repeated experimentation.

Title Page from *The Fabric of the Human Body* by Andreas Vesalius.

Illustrations from *The Fabric of the Human Body* by Vesalius showing human musculature.

THE RISE OF MODERN BIOLOGY

Circulation
Microscopes and Cells
Photosynthesis
Physiology
Nutrition
Hormones

Circulation

Born in England, William Harvey (1578-1657) studied medicine at the University of Padua in Italy. Here he was taught Galen's conception of the movement of blood which involved a special substance passing from the intestines to the liver which converted this substance into the blood which flows through the veins. Harvey's knowledge of the anatomy of the blood vessels and heart, gained from the dissection of forty different species of animals, and his observations concerning the quantity of blood present and the time required for its passage through the body, caused him to question Galen's ideas. He reasoned that if the capacity of the left ventricle was two ounces and if the heart beats seventy-two times in one minute, then in one hour the left ventricle would be sending into the aorta 72 x 60 x 2 or 8,640 ounces (540 pounds) of blood. "Not deeming it possible for the digested food mass to furnish such an abundance of blood without totally draining the veins or rupturing the arteries unless it somehow got back to the veins from the arteries and returned to the right ventricles of the heart," he wrote in 1628, "I began to think that there was sort of motion as in a circle."* Here was the idea of the same blood continuously circulating throughout the body in a closed system of tubes.

Although Harvey did not demonstrate the capillaries, his work made their existence a logical necessity. Marcello Malpighi (1628-1694) was among the first scientists to use microscopes to study animal and plant structure. While microscopically observing frog lung tissue, he recorded in 1661 that "so great is the branching of these vessels as they go out, here from a vein, there from an artery, that order is no longer preserved, but a network appears made up of the prolongations of both vessels."† Now with this first description of capillaries, Harvey's account of the circulation was complete.

*Anatomical Dissertation Concerning the Motion of the Heart and Blood In Animals—William Harvey (1628). In *Works*, Robert Willis, Editor. Sydenham Society. London 1847.

†On the Lungs—Marcello Malpighi (1661). In *Proceedings of the Royal Society of Medicine*. Vol. 23. James Young, Translator. London. 1929.

William Harvey

Title Page from *Anatomical Dissertation Concerning the Motion of the Heart and Blood in Animals* by William Harvey.

Microscopes and Cells

Robert Hooke (1635-1703) improved upon earlier microscopes by using two lenses in combination. With these compound microscopes he observed individual cells in thin sections of cork and estimated their size and the number of them that were present. He recorded these observations in 1665.

Anton van Leeuwenhoek (1632-1723) pursued as a hobby the construction of microscopes and the examination of different materials with them in the spare time from his position as a Dutch city official. He ground the lenses himself making more than four hundred, some with magnifications of more than two hundred fifty times. As a research worker, Leeuwenhoek was self-taught, never having received formal scientific training. Yet he took careful notes of all that he observed and sent them to the Royal Society of London in the form of letters. Leeuwenhoek was a keen observer, having described human sperm, bacteria, protozoa, and the capillary circulation and although he coined no new scientific terms and attempted no classification, his recorded observations of the microscopic world stimulated future investigation by microscopists.

More than two hundred years after Robert Hooke's description of the cell, Robert Brown (1773-1858) reported the presence of the nucleus in plant cells. His observations on the nucleus stand unmodified and without correction.

In the 1960's biologists began using the scanning electron microscope to study in three dimensions the fine structure of specimens. With a range of magnification from about 30 to 50,000 times, this microscope can differentiate between objects as small as 400 billionths of an inch, the size of a molecule. Since the scanning microscope has the greatest depth of focus of any microscope available, at any given

Anton van Leeuwenhoek

magnification just about everything in the field is in focus. One of the greatest advantages of the microscope is that specimens could be viewed in their entirety, there being no need to cut them into less than paper-thin slices.

Photosynthesis

Because the Belgian physician Jean Baptiste van Helmont (1577-1644) thought that water was a major ingredient of all substances, he designed the following experiment to determine the role of water in plant growth:

"I took an earthenware pot, placed in it 200 pounds of earth dried in an oven, soaked this with water, and planted in it a willow shoot weighing 5 pounds. After five years had passed, the tree grown therefrom weighed 169 pounds and about 3 ounces. But the earthenware pot was constantly wet only with rain or (when necessary) distilled water; and it was ample in size and imbedded in the ground; and to prevent dust flying around from mixing with the earth, the rim of the pot was kept covered with an iron plate coated with tin and pierced with many holes. I did not compute the weight of the deciduous leaves of the four autumns. Finally, I again dried the earth of the pot, and it was found to be the same 200 pounds minus about two ounces. Therefore, 164 pounds of wood, bark, and root had arisen from the water alone."*

This simple experiment is a model of scientific methodology. Van Helmont's approach was very quantitative

*By Experiment that All Vegetable Matter Is Totally And Materially of Water Alone—Jean Baptiste van Helmont. In *Ortus Medicinae*. Published by his son Franz M. van Helmont. Amsterdam. 1648. English Translation, *Physick Refined*, John Chandler. London. 1662.

and among the measures that he took to increase the accuracy of his experiment were drying the earth in an oven before and after weighing, using only rain or distilled water, and covering the pot in order to prevent dust from mixing with the earth.

Joseph Priestley (1733-1804) was a British scientist who emigrated to the United States in order to escape from the political bitterness generated by the French revolution. In 1774 he focused sunbeams on a red oxide of mercury causing the release of oxygen. Priestley, however, called the gas "de-phlogisticated air" because the dominant theory of chemistry at the time held that all substances contained a kind of fiery principal known as phlogiston which was given up when something burned. Just as van Helmont linked water to plant growth, Priestley explored the relationship of atmospheric gases to plants. He demonstrated that plants restored oxygen to air from which the oxygen had been removed by burning. His experiments were simple in design but well controlled, as shown by the following example. He wrote, "I put a sprig of mint into a quantity of air, in which a wax candle had burned out, and found that, on the 27th day of the same month, another candle burned perfectly well in it. This experiment I repeated, without the least variation in the event, not less than eight or ten times in the remainder of the summer. Several times I divided the quantity of air in which the candle had burned out, into two parts, and putting the plant into one of them, left the other in the same exposure, contained also, in a glass vessel immersed in water, but without any plant; and never failed to find, that a candle would burn in the former, but not in the latter."*

*Observations On Different Kinds Of Air—Joseph Priestley. In *Philosophical Transactions of the Royal Society*. London. Vol. 62. 1772.

Priestley's work made it clear that a balance between plant and animal activities was responsible for maintaining an atmosphere capable of supporting life. This important concept met with immediate widespread acceptance and stimulated a great deal of interest in photosynthesis.

During the summer of 1778, the Dutch physician Jan Ingenhousz (1730-1799) performed over five hundred experiments with green plants. He extended Priestley's discovery that green plants release oxygen by demonstrating that this can occur only in the presence of sunlight. One of his experiments involved doing the following: "Two handfuls of leaves of French beans were put in a jar of a gallon; it was kept inverted upon a dish, and some water poured upon it; next morning I found the air so much fouled that a candle could not burn in it. After having taken out some of the air for trial, I placed the jar with the remaining air and leaves in the sun from nine till eleven o'clock, when I found the air so much mended, that a candle could burn in it. After this I replaced it again in the sun till five in the afternoon, when I found the air so much mended as to be equal in goodness to common air."*

Physiology

William Beaumont (1785-1853) was an army surgeon at Fort Mackinac in Michigan. In 1822 a young Canadian accidentally suffered from a gun shot wound that left a one-half-inch opening in his stomach with lung and stomach

---

Experiments Upon Vegetables—Jan Ingenhousz. London. 1779.

tissue protruding to his body's outer surface. A small fold of tissue formed at this opening and protruded until it eventually filled the aperture eliminating the need for compresses or bandages in order to cover it. Beaumont noted that this valvular formation provided direct access into the wounded man's stomach and saw this as an opportunity to study gastric digestion. After persuading the man to remain with him for about ten years, Beaumont carried on a series of investigations that revealed the length of time for, and the manner of digestion of various foods, the effects of exercise, weather and emotions on the digestive process, the structure of the lining of the stomach and the nature and action of gastric juice.

Claude Bernard (1813-1878) was a professor of physiology at the Sorbonne where he earned a reputation as a great experimenter. He made very concentrated studies of the metabolic role of the liver and characterized the organ as a vital laboratory. In addition to his extensive analysis of liver function, Bernard also developed the concept and term of internal secretion, described the functions of the pancreatic juice, demonstrated that the sympathetic nervous system controlled the size of the blood vessels, and discovered the effects of many drugs and poisons on the body including the mechanism of carbon monoxide poisoning. The following paragraphs describing Bernard's attempt to discover the process by which sugar is formed in the liver provide insights into his experimental style:

"I chose an adult dog, vigorous and in good health, which had been fed for several days exclusively on meat, and I sacrificed it by severing the medulla seven hours after an ample meal of tripe. The abdomen was immediately opened; the liver was removed avoiding injury to its tissue, and the organ while still warm and before the blood had had time to coagulate in its vessels, was washed with cold water through the portal vein. Under the influence of this energetic washing, the liver swelled, the color of its tissue became pale, and the blood was expelled with the water which escaped in a strong continuous jet through the hepatic veins. At the end of a quarter of an hour the tissue of the liver was already nearly bloodless, and the water which emerged from the hepatic veins was entirely colorless. I subjected the liver to this continuous washing for 40 minutes without interruption. I had determined at the beginning of the experiment that the red colored water which flowed out of the hepatic veins was sweet and gave an abundant precipitate on heating, and I verified at the end of the treatment that the perfectly colorless water which emerged from the hepatic veins did not contain any traces of sugar.

The liver was then removed from the action of the water current; and I made sure, by boiling a piece of the liver with a little water, that its tissue was well washed, since it no longer contained sugar. I then left this liver in a jar at room temperature. After twenty-four hours I found that this organ, washed entirely bloodless, which I had left the night before completely free of sugar, now contained sugar in abundance. I was sufficiently convinced of this when I examined a little of the liquid which had flowed out around the liver and which was very sweet; then by injecting cold water into the portal vein with a small syringe and collecting the water as it escaped through the hepatic veins, I found that this liquid fermented very abundantly and very actively with yeast. This simple experiment, in which one can see before his eyes the abundant reappearance of sugar in a liver which had been completely deprived of it and of its blood by means of the washing, is most instructive for the solution of the problem of glycogen function. This experiment clearly proves, as we have already said, that in a fresh liver in the physiological, i.e., functional, state there are two substances: (1) sugar, which is very soluble in water and is carried away by the blood during washing; and (2) another substance, so little soluble in water that it remains bound to the hepatic tissue after the latter has been freed of its sugar and its blood by forty minutes' washing.

It is the latter substance which, in the undisturbed liver, gradually changed to sugar by a kind of fermentation, as we shall show."*

Nutrition

Hippocrates in the fifth century B.C. described the symptoms of a disease occurring among soldiers that was scurvy. The disease was also definitely described during the thirteenth century in the records of the Crusades. Before extended sea voyages became common, scurvy appeared wherever people lived on restricted diets such as in soldiers' camps and in besieged towns. During the sixteenth, seventeenth, and eighteenth centuries, when long voyages of exploration were taken, scurvy became so prevalent that it was called the "calamity of sailors." Sea air or salted meat was thought to be its cause. The naval surgeon James Lind (1716-1794) in 1747 recommended that lemon juice be given to the men on long sea trips and this dietary supplement was so successful in preventing the development of

---

*On the Mechanism of Formation of Sugar In the Liver. Claude Bernard. In *Comptes—Rendus de l'Acadèmie de Sciences*. Vol. 41. Paris. 1855.

scurvy that in 1795 lemon juice was made a compulsory component of the diet of all ships of the British Navy.

In 1895 Christian Eijkman (1858-1930) was appointed to a commission sent by the Dutch government to its East Indian colonies in order to determine the cause of beriberi, a disease that was crippling the native population. Since at the time Pasteur's germ theory of disease was gaining great popularity, it was natural for this commission to seek a germ as the cause of beriberi. This was done for two years without success. As other possible factors in the incidence of the disease, Eijkman studied the effects of the age, variety, and place of origin of the rice used for food and also of the adequacy of ventilation, population density, and age of the buildings inhabited by beriberi victims, getting negative results in all cases. He finally linked the disease to the main element in the diet of the affected persons, having found the greatest incidence of beriberi among those whose diet consisted exclusively of polished rice, grain from which the outer covering had been completely removed.

Eijkman happened to notice that many chickens kept in the hospital yard showed the symptoms of beriberi victims. To explain this, he suggested that a germ was being transmitted from the beriberi patients to the chickens with the patients' left-over rice which was routinely used to feed the chickens. Subsequently, the hospital superintendent, objecting to the use of costly polished rice as chicken feed, substituted the cheaper unpolished rice. This change produced a remarkable recovery among the chickens afflicted with beriberi and made it necessary for Eijkman to revise his original hypothesis. He now concluded that the polished rice contained a beriberi-causing toxin which was neutralized by a substance contained within the grain's outer covering.

Frederick G. Hopkins (1861-1947) was one of the first to

recognize the presence in food of the substances we now call vitamins. In 1906 he fed rats a diet consisting of pure organic compounds and observed that they became unhealthy and developed deficiency diseases. The addition of small quantities of butter fat to the diet had a curative effect. The amount was so small to be negligible in terms of the proteins, fats, or oils that it was providing. Hopkins realized that the butter fat contained substances that had a definite effect on the health of the rats and he referred to these as "accessory food factors."

Casimir Funk (1884-1967) joined the Lister Institute of Preventive Medicine in London in 1910 where he began research on the cause of beriberi. He found that there was an important nutritional substance present in the outer coating of rice that was not present in the polished rice itself. This substance was not easy to find since a ton of rice polishings yielded only one sixth of an ounce. Finally, Funk isolated a small crystalline bit of the rice husk that was capable of curing beriberi in pigeons. This was what later would be known as thiamine or vitamin B_1.

In 1912 Funk published a paper, "The Etiology of Deficiency Diseases," which stated that substances called "vitamines" were essential to life. He coined the name from "vita" the Latin term for life and "amine" the word meaning nitrogen-containing compound. Later, when it was found that not all vitamins contained nitrogen, the "e" was dropped.

Funk's finding that a definite chemical material was capable of preventing a disease changed the thinking of the day. Medicine at the time was dominated by Pasteur's work which concentrated on the effects of germs. Funk helped to establish vitamin deficiencies as the causes of diseases like scurvy, beriberi and rickets.

In 1916 pellagra was a common disease in the southern

United States believed to be caused by an insect-borne parasite. After visiting hospitals and orphanages where the disease was prevalent, Joseph Goldberger (1874-1929) observed that nurses and attendants weren't affected. This led him to suspect a dietary relationship. Twelve volunteer convicts were put on a restricted diet and after several weeks seven developed pellagra. A change of diet cured all of them and Goldberger called the effective factor P-P for pellagra preventive. After his death it was called vitamin G in his honor and today is known as niacin.

In 1934 Henrik Dam (1895-1976) reported the first account of the role that vitamin K played in coagulating blood. Having observed that chicks he was using in laboratory work developed a tendency to internal bleeding when maintained on a certain diet, Dam conducted experiments to find the dietary ingredient that restored normal blood clotting. This element he designated as vitamin K, from the German word "koagulation." Dam's researches were subsequently applied to tests on other animals and in 1944 he reported that clinical tests on humans suffering from hemorrhage showed that vitamin K helped to check the flow of blood.

Hormones

Isolated knowledge about endocrine glands could be traced back to the writings of the ancient Greeks and Romans that contain references to goiters but the first major achievement of clinical endocrinology occurred in 1855 when Thomas Addison (1793-1860) described the syndrome associated with the deterioration of the human adrenal cortex.

Edward C. Kendall (1886-1972) was attempting to isolate the secretion of the thyroid gland when alcohol containing some thyroid products that he was working with evaporated. From the white residue that remained after this evaporation, Kendall in 1914 obtained the first crystals of the thyroid hormone. In 1917 the cost of producing a gram of this hormone was $350 but ten years later the hormone was successfully synthesized.

Walter B. Cannon (1871-1945) used X-rays to observe and study peristalsis in cats. He found that injection of small amounts of adrenalin inhibited peristalsis by diverting blood from the digestive tract.

In 1872 Claude Bernard (1813-1878) proposed that substances present in the intestine had a stimulating effect on the pancreas which he attributed to a nervous mechanism.

In 1902 William M. Bayliss (1860-1924) and Ernest H. Starling (1866-1927) investigated another possible mechanism of pancreatic secretion. They completely cut the nerve supply of a loop of the jejunum so that it was connected to the body of a dog only by its arteries and veins. Hydrochloric acid was introduced into both the intact duodenum and this enervated part of the jejunum, and a cannula was inserted into the pancreatic duct in order to collect pancreatic juice. They found that the introduction of acid into either section produced exactly the same effect, the secretion of pancreatic juice. Since the loop of jejunum was completely cut off from its nervous connection with the pancreas, they concluded that a chemical substance carried by the blood to the pancreatic cells stimulated the secretion of the pancreatic juice. A piece of jejunum was then ground up and mixed with hydrochloric acid. After being filtered, this extract was injected into an animal's vein and observed to cause a flow of pancreatic juice. Bayliss and Starling suggested the name "secretin" for the active principle in the extract and with their work the first chemical substance recognized as a hormone was isolated.

Knowledge about diabetes mellitus can be traced back to records from the early days of the Roman Empire. Years later, in 1657 a close friend of William Harvey indicated that he had tasted the urine of a diabetic patient and found it to be sweet. A very significant development in this knowledge occurred in 1869 when Paul Langerhans (1847-1881) discovered that besides cells in the pancreas that secrete digestive juice, there were other cells present, the functions of which he could not determine. By 1888 Joseph von Mering (1849-1908) and Hermann Minkowski (1864-1909) determined that surgical removal of the pancreas resulted in severe and fatal diabetes. However, it wasn't until 1921 that Frederick Banting (1891-1941) and Charles Best (1899-   ) completed further investigations into the nature of the relationship between diabetes and the pancreas. They attempted to prepare an extract from pancreatic tissue that contained what they called its internal secretion. To accomplish this they gave prolonged injections of secretin to dogs in order to stimulate the pancreas to release its pancreatic juice. This was necessary in order to prepare an extract from the gland that would not be affected by the pancreatic proteases. The glands, which now did not contain pancreatic juice because of the secretin injections, were than removed and macerated and filtered. When the extracts that resulted from this procedure were injected into healthy animals, a reduction in the blood sugar level was observed.

Few advances have had as great an immediate impact on medicine as did the discovery of insulin, which was the active principle in Banting's and Best's extract. The drug dramatically saved the lives of many diabetics who were withering to death on starvation diets, the only form of treatment that doctors then had for the disease. The practice of medicine was revolutionized because patients were taught how to administer their own insulin injections. In addition, other investigators were sparked by insulin's discovery to find more of the body's hormones and new methods of treatment for patients with glandular disorders.

In 1954 Frederick Sanger (1918-   ), after ten years of intensive research, succeeded in describing the molecular structure of insulin making it the first protein for which a structural formula could be written.

Ramifications of the rapidly expanding area of endocrinology have recently extended to such widely diverse fields as malignancy, growth and aging, personality development, commercial milk, beef and fowl production, and weed control. The impact of the hormone concept on society is clearly evidenced by the fact that the chemical industry concerned with the isolation, synthesis, distribution and sale of hormones represents a multimillion dollar aspect of our economy.

By correctly piecing together the 777 atoms forming human insulin, Panayotis B. Katsoyannis (1924-   ) in 1966 achieved the first synthesis of a protein. This was not an easy task since insulin is formed from two chains, one with twenty-one amino acids, the other with thirty, held together by paired sulfur atoms. Then in 1969 Dorothy C. Hodgkin (1910-   ), using X-ray crystallography, deciphered its three dimensional structure.

The human growth hormone was synthesized for the first time in 1971 by a group of basic research scientists at the University of California Medical Center led by Choh Hao Li (1913-   ). It was the largest molecule ever pieced together by scientists in a laboratory and required 188 individual amino acids which first had to be identified and then placed in the proper order. In the synthesis, tiny polystyrene beads were used as foundations on which to build up each molecule. It was possible, by chemical manipulation, to fix the first amino acid of the sequence to each bead. The mixture was then washed clean of residual chemicals and a new process begun to affix the next amino acid.

THE CONQUEST OF DISEASE

Vaccination to Antibiotics
Virology
The Search For A Polio Vaccine

Vaccination to Antibiotics

Disease has always been a part of life. Ancient man regarded disease as the work of demons while the Greeks developed legends to explain its origin. About 500 B.C. the humoral theory of disease was developed and it dominated the practice of medicine for nearly two thousand years. According to this theory, a proper balance of four humors or fluids—blood, phlegm, yellow bile and black bile—was required for good health. Disease resulted when this balance was disturbed. Treatment required restoration of the proper balance of the humors. The leading Greek physician Hippocrates (460-377 B.C.) believed that nature would restore this balance if the patient was given rest, food and care. Other practitioners of the humoral theory thought that the balance could be restored by blood-letting. This practice persisted for centuries and even George Washington was bled with leeches during his terminal illness.

One of the earliest biologists to question the validity of these ancient theories of the cause of disease was Fracastoro of Verona (1483-1553), who suggested that diseases might be due to invisible organisms transmitted from one person to another. The development of this concept was a big step forward along the road to preventing and treating infectious diseases.

Edward Jenner (1749-1823)

Edward Jenner practiced as a physician in the English countryside where he noted variations in the ability of different groups of people to resist smallpox. Dairymen and milkmaids never seemed to contract the disease. Country people believed in the immunizing effects of cowpox. Jenner set out to test this "old wives' tale" by introducing material taken from sores on the hand of a dairymaid who had contracted cowpox into the arm of a healthy boy. About a week later, this boy developed a mild illness from which he rapidly recovered. Nearly a month later, Jenner inoculated him with material taken from smallpox sores and no symptoms of the disease were observed.

Today we would consider cowpox virus to be a mutant of smallpox virus which has lost some of its virulence for man but which still retains the ability to induce formation of virus-neutralizing antibodies in man.

In 1803, five years after the publication of Jenner's report, the annual average of deaths from smallpox dropped from 2000 to 600 in London alone. So effective has the method originally developed by Jenner been in virtually eradicating smallpox, that in this country standard vaccination for the disease has recently been discontinued. Officials of the World Health Organization expect smallpox to be eliminated from the world in a very short time.

Louis Pasteur (1822-1895)

Louis Pasteur was drawn to an investigation of fermentation because tradesmen supplying milk and wine frequently found their products to be "diseased" due to problems in controlling the process. His microscopic examination of fermenting solutions revealed the involvement of microorganisms. Pasteur demonstrated a method of controlling fermentation through the use of heat and this process, later called "pasteurization," has also been used to eliminate the risk of diseases from microorganisms present in milk.

In 1865 Pasteur was called to the south of France to deal with a disease of silkworms which was ruining the French silk industry. After three years of investigation, Pasteur succeeded in isolating the microorganisms that produced the disease and in demonstrating a method for their control. This discovery saved not only the French silk industry but also those of all silk producing countries.

After returning to Paris, Pasteur directed his attention to anthrax, a disease which was affecting sheep, and to cholera, a disease of chickens. For each of these diseases he isolated the causative microorganisms and also developed specific immunization methods which he called vaccination in acknowledgement of the influence of Jenner and his success in producing immunity to smallpox.

In 1885 Pasteur published the results of his work on a method to prevent rabies that involved injecting dogs with

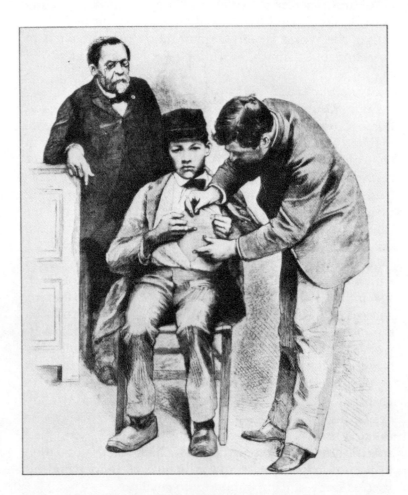

Louis Pasteur observing a doctor administer the new rabies serum he had just developed to a boy bitten by a rabid dog.

pieces of dried spinal cords taken from rabbits that had the disease. The spinal cords were dried for varying periods from fourteen days to one day, with those dried for the longer time being used first. This drying process served to attenuate the rabies virus present in the spinal cord. The injected dogs developed immunity to rabies and could even be inoculated with rabies virus directly in the brain without any subsequent development of the disease.

Pasteur was still experimenting with this method on animals when a nine-year-old boy who had been badly bitten by a rabid dog was brought into his laboratory. Pasteur administered the series of injections of dried spinal cord material and successfully prevented rabies.

Scientists are now trying to develop a rabies vaccine that can accomplish in three or four relatively painless shots what the currently used vaccine, which is similar in principle to the one originally developed by Pasteur, takes fourteen to twenty-one potentially pain-ridden injections to do.

Ignaz Semmelweiss (1818-1865)

In Paris during the nineteenth century, one in every nineteen pregnant women who went to hospitals to have babies died of puerperal (child-bed) fever. Ignaz Semmelweiss was an obstetrician in a Viennese hospital where he noticed that this disease was less prevalent in the wards attended by midwives than in the wards covered by medical students. Once, when he encountered a crying woman awaiting admission to the hospital, his inquiry revealed that she believed that her assignment to the students' wards meant certain death.

In 1847 Semmelweiss was participating in an autopsy on the body of a friend when he observed conditions in the body similar to those he had seen in the bodies of women who had died of puerperal fever. An explanation for the

Ignatz Phillip Semmelweiss

high death rate in the student ward became clear; the students attended postmortems, the midwives did not. The students brought the causative microorganisms directly from the autopsy room to the obstetric ward and examined women in labor without even washing their hands. Semmelweiss instituted the practice of washing and scrubbing in chlorine water before attending a patient and the mortality from puerperal fever dropped.

His recommendation was not readily accepted and the opposition to his views from the foremost obstetricians of the time was so fierce that Semmelweiss was forced to resign his hospital position. Subsequently though, his theories became generally accepted and were put into universal practice.

Joesph Lister (1827-1912)

In the 1860's there was very little chance of a patient surviving surgery because of the fatal effects of infections that were contracted during the operation. It is estimated that four out of every five operations ended with the death of the patient. In fact, there was one doctor in England who was reported to have performed one thousand operations with only three of his patients surviving. Since operating rooms of that period usually had sawdust on the floor to absorb blood and the surgeons often wore gowns covered with dried blood and pus, one can understand the existence of such a low surgical success level.

Joseph Lister was familiar with Pasteur's research that demonstrated the role of microorganisms from the air in causing decomposition and disease. He applied Pasteur's findings to the treatment of wounds and surgery, believing that the microorganisms from the air or from the surgeon's hands and instruments were causing the infections which resulted in the deaths of so many patients.

Lister attempted to prevent these infections through the use of chemicals that would prevent the growth of the microorganisms but at the same time would not injure human tissue. He put carbolic or phenic acid into the wounds accompanying compound fractures and thus introduced the use of antiseptics. Later, he devised methods of controlling microorganisms in the operating room by spraying carbolic acid over the hands of the doctor before an operation and on the immediate surroundings while the operation was in progress.

The reluctance of the medical profession to adopt Lister's methods was reduced when in only three years the surgical death rate dropped to 15%. Today, instead of the antiseptic surgery introduced by Lister, aseptic surgery is practiced. With it, the field of operation, the instruments, and the dressings are all rendered free of microorganisms by the process of sterilization.

Robert Koch (1843-1910)

In 1876 Robert Koch discovered that the disease anthrax, which seriously affected cattle, was caused by rod-shaped bacteria. Seven years later, he developed a method of preventive innoculation against it.

Many techniques involved in the investigation of bacteria were developed by Koch. He was the first to isolate pure strains of bacteria and to grow them on a culture medium solidified with gelatin. At first he used flat glass plates to hold the culture medium but the warm gelatin ran off the edges. To prevent this, Koch's assistant, Richard J. Petri, invented a dish with the edges turned up. Koch also perfected methods of staining bacteria with dyes to facilitate greater visibility and was able to photograph them through his microscope.

In describing the bacillus that causes tuberculosis, Koch

outlined the steps or postulates that should be followed in order to prove that a specific kind of microorganism is the cause of a particular disease. It was necessary to find the suspected microorganisms in the bodies of diseased animals, to isolate the organisms from the bodies of these animals and grow them in pure culture, to cause the disease in healthy animals by injecting them with the isolated organisms from the pure culture, and to reisolate the same organisms from these artificially infected animals.

Koch's statement of these rules, together with the technical procedures introduced by him and his colleagues, paved the way for the search for the causes of infectious diseases and resulted in the isolation of about twenty of them within the last two decades of the nineteenth century. More recently, it was through the application of Koch's method that a newly discovered bacterium was proven to be the cause of the Philadelphia legionnaire's disease which killed twenty-nine persons and sickened one hundred and eighty during the summer of 1976.

Walter Reed (1851-1902)

Walter Reed was sent by the United States Government to Cuba in 1900 to investigate the cause of yellow fever which had resulted in more deaths among soldiers during the Spanish-American War than the battle itself.

To determine the role of contaminated articles in the spread of the disease, Reed had volunteers inhabit an "Infected Clothing Building" where they used bed linens and personal articles of yellow fever victims without contracting the disease.

His next approach was to establish an "Infected Mosquito Building" which was devided into two compartments by a wire screen partition. Volunteers occupied both compartments but mosquitoes infected with yellow fever were introduced into only one. The inhabitants of the side harboring the mosquitoes developed yellow fever after receiving a number of bites while the controls on the mosquito-free side remained unaffected.

As a result of Reed's discovery of the role of mosquitoes in the transmission of yellow fever, the road was now clear to eliminate this disease through the control of the mosquitoes. This was done within two years at the Canal Zone making possible the construction of the Panama Canal.

Paul Ehrlich (1854-1915)

Paul Ehrlich studied immunology at Robert Koch's Institute for Infectious Diseases in Berlin. He conceived of antibodies as "magic bullets" which seek out and kill germs that enter the body and sought to find chemical agents that would function in a manner similar to that of the antibodies. He thought that if different tissues had varying affinities for different stains, they also might have different affinities for other chemicals. Thus, certain materials might be found which would have an immediate toxic effect upon invading microorganisms without seriously affecting the tissues of the host.

In 1910, Ehrlich prepared a compound of the element arsenic called salvarsan which proved immediately effective in treating syphilis. As it was the 606th compound that he tested before achieving success, it was known as "606." Salvarsan was one of the first great therapeutic agents discovered in modern times and with it began the era of chemotherapy.

Alexander Fleming (1881-1955)

The element of chance has played a part in scientific discovery more than once. However, Louis Pasteur once

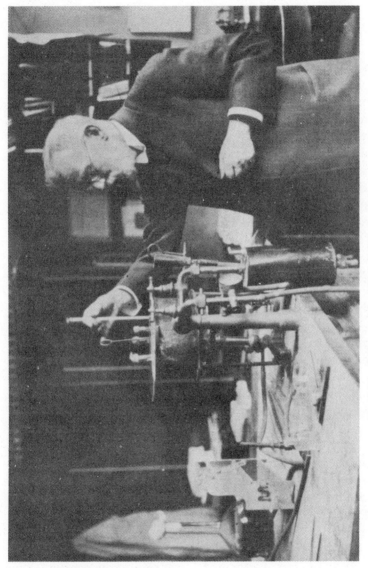

Paul Ehrlich working in his laboratory.

remarked that "Chance favors the prepared mind." Probably few scientists had more accidents or were more ready for them than Alexander Fleming.

In 1922 Fleming was working in a London laboratory while he had a severe cold. A drop from his runny nose fell on a glass plate where bacteria were being grown. Within a short time, around the region where his nasal secretion had fallen, there was a clear space devoid of bacteria. A substance in the secretion had produced this. Fleming identified the substance as lysozyme and went on to find it also in human tears, saliva, and mother's milk. In the intervening years, lysozyme has been isolated, purified, and chemically identified as a huge protein and digestive enzyme with the ability to dissolve bacteria. Recently, scientists are investigating the role lysozyme might play in controlling cancer by destroying cancer cells in a manner similar to the way in which it destroys bacteria.

While working with staphylococci, Fleming had many culture plates set aside on the laboratory bench which were examined periodically. In these examinations, the plates were necessarily exposed to the air and some became contaminated with various microorganisms. Around the colonies of a contaminating mold, Fleming observed that the staphylococcus colonies became transparent and were obviously being destroyed. He subcultured this mold and subsequent experiments revealed that the broth in which the mold had been grown had acquired the power to destroy many of the more common pathogenic bacteria. He called the antibacterial substance present in the broth penicillin.

Fleming published his results in 1929 but they created no excitement. He abandoned his experiments on penicillin three years after discovering it because his attempts to isolate it proved unsuccessful.

Howard Florey (1898-1968)

In 1938 Howard Florey stumbled on Fleming's paper and was stimulated by it to obtain a sample of Fleming's mold and begin work. A research team assembled by him grew the penicillium mold in hundreds of flasks and after months of labor was able to extract from the mold broth a tiny amount of yellowish powder. When flecks of this powder were dropped on culture plates with pathogenic bacteria, the results were spectacular.

After many tests on animals, the breakthrough came in 1934 when eighteen patients with serious staphylococcus infections were successfully treated. Faced with a lack of the funds necessary to develop penicillin in Great Britain, Florey turned to drug companies in the United States for financial support. Here, the Florey team set in to motion the mass methods necessary to produce penicillin commercially. So successful was the American effort that by the time of the Normandy invasion during World War II (1944) there was enough penicillin on hand to treat all casualties.

Penicillin became known as the "wonder drug," and after it was realized that in the course of evolution strains of microorganisms resistant to its effects were developing, the search for other antibacterial substances began.

Selman Waksman (1888-1973)

In 1941 Selman Waksman coined the term "antibiotic" to describe penicillin. He defined it as a "chemical substance produced by a microbe which has the capacity to inhibit the growth of and even to destroy other microbes."

Soon after writing that definition, Waksman isolated and tested 10,000 microbes for their ability to inhibit the growth of pathogenic bacteria. He found that ten percent lent themselves to the process. From these one hundred he tried to extract the active substances and ended up with ten chemical compounds. These were tested in animals and one was the antibiotic now known as streptomycin, the first such substance to be effective in treating tuberculosis.

Millions of dollars in royalties from streptomycin and other subsequently discovered antibiotics such as neomycin were used to build and operate the Rutgers Institute of Microbiology which was founded in 1949 and headed by Waksman until he retired in 1958.

Virology

Virology was a relatively new science in the 1930's starting out as an extension of bacteriology. Concepts of viruses varied from their being considered small bacteria to poisonous liquids.

In 1898 Dimitri Iwanowski (1864-1920) studied a disease of tobacco plants called tobacco mosaic disease. He extracted from diseased tobacco leaves a fluid that contained the organism that caused the disease. This fluid was passed through a fine porcelain filter that prevented the passage of the smallest known bacteria. Since the organism that caused tobacco mosaic disease passed right through this filter, he called it a filterable virus.

From a ton of infected tobacco leaves, Wendell Stanley (1904-   ) in 1935 extracted a tablespoon of a white crystalline powder. When this powder was rubbed on a tobacco leaf, the symptoms of tobacco mosaic disease appeared. Thus Stanley demonstrated that viruses assumed the properties of living things only when they were inside a living cell. Outside of the living cell, viruses could crystallize like organic salts.

In 1910 Francis P. Rous (1879-   ) was working with a soft tumor known as a sarcoma in chickens. He was seeking some way to transmit the tumors from one chicken to another. Rous ground the tumors to the consistency of a soup and then strained the soup through fine filters, carefully extracting all of the cells and any germs that he could find. He then injected the remaining clear fluid into ten chickens and observed the development of tumors in all of them. This was the first report of a virus causing malignant growths of the sarcoma type in chickens. Subsequently this Rous virus was found to evoke tumors in a large number of animal species including mammals. The significance of Rous' work was realized with the isolation of the leukemia virus in mice in 1951.

The Search for a Polio Vaccine

During the 1940's virology became a fast-growing science with two rival schools of research. There were the M.D.'s who were concerned with the different diseases viruses might cause and the Ph.D.'s who were concerned with the actual nature of viruses.

At this time, poliomyelitis (infantile paralysis) was one of the great epidemic diseases. In 1908 Karl Landsteiner (1868-1943) demonstrated that it was caused by a filterable virus. Thus much of the virus research was directed towards the development of a vaccine to prevent it. This direction was also encouraged by the establishment of the National Foundation for Infantile Paralysis which collected contributions from the public and supported the work of scientists endeavoring to produce an effective vaccine against the disease.

Jonas Salk (1914-   ) and Thomas Francis (1900-1969) had developed a killed virus flu vaccine and were familiar with some of the problems that would be encountered in the development of one for polio. These related to the amount of formalin necessary to kill the virus, the selection of the specific types of virus to be used, and the possibility that a strain might show low virulence in repeated trials and then revert to virulence when injected as a vaccine.

John Enders (1897-   ) successfully grew mumps virus in tissue culture made from human embryonic tissue obtained from stillborn babies. When his experiments were completed he observed that polio virus was also able to grow on some remaining human embryonic tissue as well as on mature human tissue and non-nervous monkey tissue. Previously it was thought that polio viruses could be cultured only in nervous tissue, a material that could not be used in vaccine production because of the danger of transmitting encephalitis with bits of infected nerve tissue.

To some workers the presence of polio antibodies in the blood seemed to indicate that the point of entry and the route that the polio virus traveled was not the nervous system although established virologists could not be swayed from their long held conception of polio as a disease of the nervous system. However, in 1951 gamma globulin was used to produce passive immunity against polio. The effectiveness of small doses of gamma globulin strongly suggested that the polio virus enters the nervous system via the bloodstream and that polio might be a widespread intestinal infection.

Before an effective polio vaccine could be developed all the known strains of polio viruses had to be typed. In 1941 Salk started doing typing using monkeys because they were susceptible to all of the three suspected types of polio virus. After confirming that there were three types and many strains, he found one strain for each type that would grow well in tissue culture and induce antibody formation after being killed with formalin. In order to determine if any live virus remained, the treated virus was injected into a monkey's brain. The subsequent development of polio indicated the presence of live virus. To determine if the treated virus was effective, it was injected intramuscularly into monkeys whose blood was then tested for the presence and quantity of polio antibodies. Experiments with monkeys were successful. But a virus harmless to monkeys could paralyze humans and antibody formation in monkeys didn't guarantee the same effect in humans. Salk believed that the time for human testing had arrived.

By March of 1954 seven thousand children had been given Salk's vaccine without any adverse reactions and all inoculated individuals had demonstrated a significant rise in the level of antibodies in the blood. There was now intense public pressure for a field trial of the vaccine although both Albert Sabin (1906-   ), who in 1935 was growing polio virus in nerve cells, and Enders thought such testing to be premature. However the National Foundation decided to proceed and the job of evaluator was given to Thomas Francis, the former teacher of Salk and early virus researcher. The largest testing of a medical product in the history of man began in April 1954 and 441,131 children received the real vaccine, 201, 229 were inoculated with a placebo, and 1,063,951 were not inoculated but observed as additional controls. Evaluation involved not only the observation of the incidence of polio among all three groups of children but also the determination of the level of antibodies in the blood and the consideration of such factors as age, geographic location, and socio-economic status. One year later Francis reported that the vaccine was found to be 60-70% effective against Type I polio virus, the most frequent cause of paralytic polio and over 90% effective against Types II and III. A few hours after the report was made the Salk vaccine was licensed for commercial production. A number of vaccine-related cases of paralytic polio were discovered and the vaccination program was briefly suspended for a check on the vaccine manufacturers. Sabin cited incidents such as this as indications that the Salk vaccine was not consistently safe and announced that he was making progress on a live virus vaccine.

In the meantime, Herald A. Cox (1907-   ) and Hilary Koprowsky (1916-   ) developed two attenuated strains of polio virus and believed that a successful large test would overcome the anxiety that people had about live virus vaccine. Because the widespread use of the Salk vaccine had raised the polio antibody level of millions of American children, such a large-scale test could not take place in the United States. Trials were conducted in Northern Ireland with the virus given in sweet syrup, but when it was discovered that vaccinated individuals were excreting large amounts of virus that had reverted to virulence, further testing was halted.

Sabin had started attenuating strains of polio virus in 1953, growing the virus, like Salk, on kidney tissue. He found and isolated mutant strains of virus that were nonvirulent but still antigenic and preliminary tests on monkeys and humans were successful. In the spring of 1957 a field trial of the Sabin vaccine was conducted in the Soviet Union with the three strains of attenuated virus suspended in syrup or candy. By 1960 fifteen million individuals in more than five countries had been successfully immunized with the oral vaccine and it was licensed for commercial production. Between 1960 and 1965 a number of vaccine-associated polio cases had been reported. For example, in 1963 eighteen cases of paralytic polio were definitely attributed to the vaccine. The revision of the manufacturing procedures has significantly reduced this number.

On the twentieth anniversary of the announcement of

the polio vaccine (1973) Salk reported that the live virus vaccine was responsible for some of the cases of polio which continue to occur in the United States. In the ten year period from 1961, after it was first licensed for use in this country, through 1971 109 vaccine associated cases were reported. Salk attributed these to the genetic instability of the virus strains used to produce the vaccine. He believes that the switch from the killed virus vaccine (his) to the oral vaccine (Sabin's) that was recommended by the American Medical Association in 1961 was ill-advised. As Salk sees it, the only advantage that could be claimed for the live virus vaccine is its oral administration and this ease of administration doesn't justify the toll of several polio cases each year. He is critical of the fact that the killed virus vaccine is neither manufactured nor distributed in this country so that people in the United States don't have the same freedom of choice that exists in other countries such as Canada and France where both live and killed virus vaccines are available. Salk maintains that his vaccine is not only effective but also completely safe and if used could completely control polio as has happened in Sweden where only his vaccine is used and where polio has not occurred since 1966.

THE ORIGIN OF LIFE

Disproving Spontaneous Generation
Returning to Spontaneous Generation

Disproving Spontaneous Generation

Until the last half of the seventeenth century, the notion that living organisms could arise from non-living matter was widely believed. According to this idea of spontaneous generation, organisms were formed from dirt, mud or decaying flesh. The Italian physician Francisco Redi (1626-1697) disagreed with those naturalists of the period who believed that Aristotle's writings about spontaneous generation should not be questioned and instead attempted to experimentally test the theory. In 1668 he wrote ". . .although it be a matter of daily observation that infinite numbers of worms are produced in dead bodies and decayed plants, I feel inclined to believe that these worms are all generated from other worms and that the decayed matter in which they are found has no other function than that of serving as a place, or suitable nest, where animals deposit their eggs at the breeding season, and in which they also find nourishment.

"Belief would be vain without the confirmation of experiment. In the middle of July, I put a snake, some fish, some eels and a slice of veal in four large, wide-mouthed flasks, having well closed and sealed them. I then filled the same number of flasks in the same way, only leaving these open. It was not long before the meat and the fish, in these second vessels, became wormy and flies were seen entering and leaving at will; but in the closed flasks I did not see a worm, though many days had passed since the dead flesh had been put in them."*

Redi's work established that macroorganisms were not generated spontaneously but belief in the spontaneous generation of microorganisms persisted. In 1765 Lazaro Spallanzani (1724-1799) became aware of the work of the English biologist John Needham (1713-1781), who reported data supporting the spontaneous generation of microorganisms. Needham had boiled broth in corked vessels and in a few days had observed them to be swarming with microorganisms. Spallanzani decided to repeat Needham's experiment but under more rigorous conditions. He wrote, "I took sixteen large and equal glass vases; four I sealed

*Experiments On The Generation of Insects—Francesco Redi (1668). In *Experiments on the Generation of Insects*. Open Court Publishing Company. Chicago, Illinois. 1909.

hermetically; four were stopped with a wooden stopper, well fitted; four with cotton; and the four last I left open. In each of the four classes of vases, were hempseed, rice, lentils, and peas. The infusions were boiled a full hour, before being put into the vases. I began the experiments 11 May and visited the vases 5 June. In each there were two kinds of microorganisms, large and small; but in the four open ones, they were so numerous and confused that the infusions, if I may use the expression, rather seemed to teem with life. In those stoppered with cotton, they were about a third more rare; still fewer in those with wooden stoppers; and much more so in those hermetically sealed.

"The number of microorganisms developed is proportional to the communication with the external air. The air either conveys germs to the infusions, or assists the expansion of those already there."*

Needham remained unconvinced by Spallanzani's work maintaining that the prolonged heating used by Spallanzani had altered the air and destroyed the "vegetative force" that were needed for the development of life.

About one hundred years after Spallanzani's work, Louis Pasteur (1832-1895) entered the controversy. He wrote, "In a glass flask I placed one of the following liquids which are extremely alterable through contact with ordinary air: yeast, water, sugared yeast water, urine, sugar beet juice, pepper water. Then I drew out the neck of the flask under a flame, so that a number of curves were produced in it. I then boiled the liquid for several minutes until steam issued

*Observations And Experiments Upon the Microorganisms of Infusions—Lazaro Spallanzani (1779). In *Tracts on the Nature of Animals and Vegetables*. Translated by J. G. Dalyell. Edinburgh. 1799.

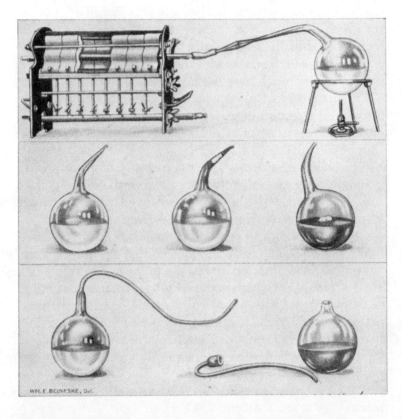

Flasks used by Pasteur to refute the theory of spontaneous generation.

freely through the extremity of the neck. This end remained open without any other precautions. The flasks were then allowed to cool. Anyone who is familiar with the delicacy of experiments concerning the so-called 'spontaneous' generation will be astounded to observe that the liquid treated in this casual manner remains indefinitely without alteration. The flask can be handled in any manner, can be transported from one place to another, can be allowed to undergo all the variations in temperature of the different seasons, the liquid does not undergo the slightest alteration. It retains its odor and flavor. In no case is there the development of microorganisms in the liquid.

"It might seem that atmospheric air, entering with force during the first moments, might come in contact with the liquid in its original crude state. This is true but it meets a liquid which is still close to the boiling point. The further entrance of air occurs much slower, and when the liquid has cooled to the point where it will not kill the germs, the entrance of air has slowed down enough so that the dust it carries which is able to act on the infusion and cause the development of microorganisms is deposited on the moist walls of the curved tube. At least, I can see no other explanation for these curious results. For, after one or more months in the incubator, if the neck of the flask is removed by a stroke of a file, without otherwise touching the flask, molds and infusoria begin to appear after 24, 36, or 48 hours, just as usual, or as if dust from the air had been inoculated into the flask."*

*On the Organized Bodies Which Exist In The Atmosphere; Examination Of The Doctrine of Spontaneous Generation—Louis Pasteur. In *Annales des Sciences Naturelles*. 4th Series. Vol. 16. 1861.

Returning to Spontaneous Generation

Although Pasteur's work signaled the downfall of the spontaneous generation hypothesis, most modern scientists approach the origin of life through just that hypothesis. It is generally believed that the formation of organic compounds played a crucial role in the development of an organism. Harold C. Urey (1893-   ) was interested in finding out what role the electrical discharges in the upper atmosphere had in the formation of organic compounds. One of his students, Stanley L. Miller (1930-   ), performed the simple experiment of circulating a mixture of water vapor, methane, ammonia and hydrogen—all gases thought to have been present in the early atmosphere of the earth—continuously for one week over an electric spark. The circulation was maintained by boiling the water in one limb of the apparatus and condensing it in the other. At the end of the week the water was analyzed by paper chromatography. It was found to contain a mixture of amino acids. The yield was high and demonstrated the spontaneous formation of amino acids, which very probably was a key step involved in the origin of life.

THE DEVELOPMENT OF GENETICS

Pre-Mendelian Inheritance
Mendel
Rediscovery
Garrod—Another Ignored Contribution
The Rise of Modern Genetics
The Genetics of Sickle Cell Anemia
Discovery of the Hereditary Material
Genetics in the Soviet Union
Synthesizing Nucleic Acids
Recombinant DNA and Genetic Engineering

Pre-Mendelian Inheritance

From the earliest times there was interest in explaining the observed resemblance of offspring to parents. Many prominent scientists offered their ideas hoping to shed light on the mechanism that made possible the transmission of traits. In 1865 Charles Darwin (1809-1882) worked out a hypothesis explaining how the characteristics of life could be transmitted to the germ cells. He called it "Pangenesis" and it is considered to be his least satisfactory effort. He believed that during the life of the organism particles from the various organs passed to the sperm and egg and through these cells parental traits were passed on to the offspring.

His cousin, Francis Galton (1822-1911), also in 1865, applied statistical procedures to biological data in his attempt to determine the proportionate contribution of each generation of ancestors to the genetic make-up of the individual. For example, he computed that to the total heritage of the individual the parents contribute one half, the grandparents one quarter, and the great-grandparents one eighth.

August Weismann (1834-1914) referred to the protoplasm of sperm and egg as germplasm and stated that it is only the germplasm that is passed on from generation to generation and through this germplasm the hereditary elements are transmitted.

Gregor Mendel

Gregor Mendel (1822-1884) received a scientific education and then entered an Augustinian monastery in the town of Brunn in Austria where he taught high school science. During his free time Mendel mated pea plants in the monastery gardens and observed and recorded the traits of the resulting offspring. He cross pollinated pure breeding plants with contrasting traits such as yellow or green seeds, round or wrinkled seeds, and red or white flowers. Mendel transferred the pollen himself, took the

Gregor Mendel

necessary measures to prevent accidental pollination, gathered and planted the resulting seeds and personally observed and recorded the traits of the offspring that developed from these seeds. He always found that the offspring from these crosses were alike and resembled only one of the two parents. When these offspring were crossed among themselves the offspring showed either of the two original contrasting traits. From breeding experiments such as these emerged the conception of the hereditary factors as separate particulate entities in the cell. Mendel concluded that the pea plant possesses two hereditary factors for each trait and that when gametes are formed the two factors segregate and pass into separate gametes so that each gamete contains only one factor for each trait. Each new plant thus receives one factor for each trait from its male parent and one for each trait from its female parent.

Mendel's conclusions are consistent with what we know about the chromosomes and their behavior in meiosis. Yet Mendel did his work before the details of cell division and fertilization had been learned and before the chromosomes and their role in heredity had been discovered. A significant contribution to Mendel's successful analysis was his introduction of several innovative procedures. He reduced the problem to its simplest terms by limiting his study to one trait at a time. In one set of crosses he would concentrate only on seed shape while in another he would observe flower color. Mendel realized that "the value and utility of any experiment are determined by the fitness of the material to the purpose for which it is used" and established criteria to guide the selection of plants for his experiments. For example, he wrote that "The experimental plants must necessarily possess constant differentiating traits (and) the hybrids of such plants must, during the flowering period, be protected from the influence of all foreign pollen . . ."* Finally, Mendel mathematically analyzed the results of his experimental crosses using a knowledge of statistics that he obtained from his extensive study of physics. What Mendel did was to apply the physicist's approach to experimentation to the science of biology.

After eight years of breeding plants Mendel published his work in 1866 in the journal of the Natural History Society of Brunn. This paper was ignored by the scientific community and soon forgotten. Several reasons have been suggested to explain why Mendel's paper went unnoticed. The journal in which it was published was a local one and not widely known and the novel ideas presented in it were not easily understood by the scientific community because the details of cell division and fertilization were not yet known. These details were completely worked out first by 1900 and then, after thirty-four years of neglect, Mendel's classic paper was simultaneously rediscovered by three investigators working independently of one another in three different countries.

Rediscovery

The Dutch biologist Hugo DeVries (1848-1935) began crossing plants in 1892 and like Mendel also obtained three-to-one ratios of dominant to recessive offspring from the mating of hybrids.

In 1900, while preparing to publish his work, he received a reprint of Mendel's papers from a friend familiar with his

*Experiments in Plant Hybridization—Gregor Mendel (1865). In *Proceedings of the Natural History Society of Brunn.* Translated by W. Bateson. "Mendel's Principles of Heredity." Cambridge, 1909.

hybrid studies who thought that it would be of interest to him. Then DeVries realized that he had been anticipated and he added references to Mendel's work to his paper.

Both Carl Correns (1864-1933) of Germany and Erich Tschermak (1871-1962) of Austria arrived at correct explanations for the three-to-one ratios they were obtaining in their breeding experiments in 1899 although each was unaware of the other's success. Similarly, each had independently discovered Mendel's paper as a result of reading the same reference book in 1900. Thus Mendel's neglected paper was simultaneously rediscovered by three different men, each of whom had independently performed experiments that led to the same conclusions Mendel had reached thirty-four years before.

Archibald Garrod (1857-1936)—Another Ignored Contribution

One of Archibald Garrod's patients produced urine that turned black upon exposure to air because of the presence in the urine of a chemical called alkapton. His study of a number of families in which this condition, alkaptonuria, appeared led him to the conclusion that it was inherited as a simple Mendelian recessive trait. In 1899 Garrod suggested that alkaptonuria resulted from the inability to produce an enzyme that catalyzes a single metabolic step. The production of the enzyme was controlled by heredity, therefore he called the disease an "inborn error of metabolism." Normal individuals produce the enzyme that catalyzes the oxidation of alkapton to carbon dioxide and water but those who have alkaptonuria lack this enzyme and must excrete alkapton which is not broken down. This was the first attempt to link heredity and biochemical reactions. So novel was this idea that it caused no excitement in the scientific community and Garrod's paper was forgotten until the work of Beadle and Tatum in 1941. Like Mendel's work, this was another example of a discovery that was made before the scientists of the time were ready for it. Its fate was to remain forgotten and ignored until scientific knowledge advanced to the point when the discovery could be clearly understood. Then, with its significance properly appreciated, it was brought out of oblivion.

The Rise of Modern Genetics

The American geneticist Thomas Hunt Morgan (1866-1945) more than any other single person brought classical genetics to a high level of development. He sought an appropriate experimental animal and out of thousands of possibilities hit upon the fruitfly *Drosophila melanogaster*. Perhaps no other organism has contributed so much to our knowledge of genetics as has this common fruitfly. Among its virtues as a subject for genetic experiments are the ease with which large numbers can be cultured in the laboratory and that there is a new generation about every twelve days with a large number of offspring produced from a single mating. Use of these flies made it possible to complete in months experiments that with other organisms might require years to complete. Morgan casually bred his flies in milk bottles with slices of bananas as food. Envelopes from his correspondence served as supports on which the flies' larvae would develop into pupae.

In 1910 Morgan presented clear-cut evidence that a specific trait in *Drosophila melanogaster*, white eyes, was linked to the inheritance of sex and most likely associated with a particular chromosome, the X. He hypothesized that genes occurred in linear order along the length of chromosomes and that linkage was a physical relationship between genes that could be modified by a physical crossover between gene pairs on homologous chromosomes.

That same year Morgan took two Columbia University undergraduates, Alfred H. Sturtevant (1891-1970) and Calvin B. Bridges (1889-1938), to work in his laboratory, and a third one, Hermann J. Muller (1890-1966), soon joined them. It was one of those rare circumstances in the development of science where an inspiring scientist and teacher, an extremely gifted group of students, and a research opportunity that was very ripe for exploitation were brought together at the same time. This was the beginning of the famous "fly room" at Columbia University. The small room had eight desks crowded into it along with incubators and racks of milk bottles containing fly colonies. From 1910 to 1927 Morgan, Sturtevant and Bridges kept the fly room in continuous operation. During this period Columbia University attracted many foreign students and a flood of papers was put out by the group. Sturtevant described the activities in the fly room in the following way: "The group worked as a unit. Each carried on his own experiments, but each knew exactly what the other was doing, and each new result was freely discussed. There was little attention paid to priority or to the source of new ideas or new interpretations. What mattered was to get ahead with the work. ... There can have been few times and places in scientific laboratories with such an atmosphere of excitement and with such a record of sustained enthusiasm." In 1928 the group moved to the California Institute of Technology and made it one of the world's leading centers of genetic research.

Sturtevant, in 1913, as a result of his study of the frequency of crossing over and its resultant gene recombinations, prepared the first chromosome map showing the locations of specific genes. Bridges, in that year, observed that nondisjunction of chromosomes during meiosis produced sperm or eggs with either one extra chromosome or with a deficiency of one chromosome, resulting in flies with defective traits.

Prior to the rediscovery of Mendel's work, DeVries theorized that hereditary changes in nature, rather than having been small and continuous, may have been large and discontinuous. He called these large changes mutations. Hermann J. Muller realized that the study of these mutations was seriously hampered by the extreme infrequency of their occurrence under ordinary conditions. He demonstrated that treatment of the fruitflies' sperm or egg cells with X-rays caused a fifteen thousand percent increase in the mutation rate and that the mutant genes produced from the effects of this radiation behaved in the manner typical of the mutant genes found in organisms generally.

Muller in 1922 brought the first stocks of *Drosophila* to the Soviet Union for use there in genetics research. In the 1930's he was invited to direct the Russian research program and spent several years as Senior Geneticist in the Institute of Genetics of the U.S.S.R. Academy of Sciences. He left in 1937 after having become very disillusioned with the political control being exercised over genetics which involved the imprisonment, banishment and execution of many leading Russian geneticists. In the years before his death Muller was engaged in an active campaign to warn the world of the potential danger to future generations from increased mutation rates induced by radiation from nuclear weapons' tests.

The Genetics of Sickle Cell Anemia

James Herrick (1861-1954) in 1910 observed on a routine slide made from the blood of a twenty-year-old black patient, red blood cells which were crescent shaped. Herrick's subsequent report of this unusual observation, which he had never seen described before, signaled the first awareness of a very novel medical occurrence although no specific diagnosis was made. However, once the report was published, other physicians observed similar crescent shaped cells in cases of anemia and the condition came to be called sickle cell anemia.

In 1949 James Neel (1915-   ) reviewed all of the literature pertaining to sickle cell anemia and found that in some individuals the tendency to produce sickle-shaped red blood cells did not have any clinical significance while in others it resulted in a severe anemia. He then tested a large number of parents of sickle cell anemia patients for the sickling tendency and found that the condition was present in all of them without exception. Neel concluded that there were three possible genotypes involved in the inheritance of sickle cell anemia. An individual could have two genes for normal hemoglobin, two genes for sickle hemoglobin, or one gene for normal hemoglobin and one gene for the sickle kind. Those with the last genotype were said to have the sickle cell trait.

In 1949 Linus Pauling (1901-   ) led a team of investigators that used electrophoresis to study both normal and sickle hemoglobin. The blood samples studied came from individuals who had each of the three possible genotypes for hemoglobin production described by Neel. They found that the hemoglobin of heterozygous individuals consisted of 40% sickle hemoglobin and 60% normal and also that there was an electrical difference between the two kinds of hemoglobin. Pauling's group concluded that sickle hemoglobin had gained two units of positive charge compared to normal hemoglobin and that this difference was due to some change in the protein part of the hemoglobin molecule.

The British physician Anthony C. Allison (1925-   ) was interested in finding out how the sickle cell gene could be maintained at such a high frequency among so many people in spite of the constant elimination of these genes through deaths from the anemia. In 1954 he suggested as an explanation the possibility that individuals with the sickle cell trait might, under certain conditions, have a selective advantage over those without the trait. Analysis of the world-wide distribution of the sickle cell trait revealed that its incidence was highest in regions where malaria was prevalent. A study of African children indicated that those with the sickle cell trait had malaria less frequently than children without the trait. Further evidence regarding the protective action of the sickle cell trait required direct observation of the effects of artificially induced malarial infection on volunteer subjects.

Two comparable groups of individuals were involved; one with the trait and one without it. The members of both groups were infected with the malarial protozoan and it was found that all but one individual in the group that did not have the sickle cell trait developed malaria while the entire group with the sickle cell trait remained unaffected. Allison concluded that individuals with the sickle cell trait would suffer from malaria less often than those without the trait. He thought that the sickle cells were less easily parasitized than normal erythrocytes because of differences in the nature of their hemoglobin. This would explain why those who were heterozygous for the sickle cell gene had a selective advantage in regions where malaria was endemic and why the sickle cell gene remained common in these areas in spite of the elimination of genes with the deaths of people from sickle cell anemia.

It remained for Vernon M. Ingram (1924-) in 1958 to resolve the problem related to the nature of the difference between normal and sickle hemoglobin. He used the protease trypsin to break the hemoglobin molecule down into twenty-eight fragments or peptides, each consisting of a small group of amino acids. Then he used a combination of electrophoresis and chromatography to produce a map or "fingerprint" of the molecule. Both normal and sickle hemoglobin molecules were subjected to this procedure and then the "fingerprints" of each were compared.

Ingram found that the normal and sickle cell peptides contained the same types of amino acids but that there was a difference in the amounts present. The normal peptide had two glutamic acid units and a single valine while the sickle cell peptide had a single glutamic acid and a double dose of valine. Finally, he determined the order of the amino acids in the peptide and established that in the sickle cell peptide a valine unit occupied the place of the usual glutamic acid. Thus the sole chemical difference between normal and sickle hemoglobin was shown to be the substitution of valine for glutamic acid at one point in the abnormal molecule. This seemingly small change of one amino acid in nearly three hundred could be fatal to the unfortunate possessor of the abnormal hemoglobin.

Discovery of the Hereditary Material

In 1869 Johann Miescher (1844-1895), a Swiss biochemist, treated cells with pepsin in order to make their proteins soluble and found that the nuclei shrink but remain essentially intact. This same nuclear material that can withstand peptic digestion was also found to behave unlike protein when treated with a variety of other reagents. It has since been named DNA.

Nucleic acids had been isolated from cell nuclei in 1897 but no attempt had been made to connect them to the process of heredity. The rediscovery of Mendel's paper had not yet occurred and very little was actually known about the relationship of the nucleus to heredity.

In 1914 Robert Feulgen (1884-1955), a German chemist, devised a method of selectively staining DNA a brilliant crimson, making it possible to easily identify it.

In 1928 Frederick Griffith (1891-   ) reported the results of experiments in which he studied the effects on mice of virulent and non-virulent strains of pneumococci. Mice survived injections of live non-virulent and dead virulent strains of the pneumococci and succumbed to injections of live virulent strains. These results were expected but the observed death of mice that followed the injection of a mixture of live non-virulent and dead virulent pneumococcus strains puzzled him. Griffith found that the dead mice contained large numbers of live virulent pneumococci. He became convinced that the live non-virulent strain had been transformed into virulent organisms by hereditary material from the dead virulent bacteria. These transformed virulent strains, when cultured, reproduced new virulent organisms.

By 1931 other workers had shown that the rodent host was not essential for bacterial transformation and that it could also occur in test tube cultures. Two years later it was shown that not even whole virulent strain cells were necessary but that live non-virulent cells in a test tube could be transformed into virulent organisms by fluid in which dead virulent bacteria had been dissolved.

The hereditary characteristics of the virulent strain appeared to have been transmitted to the cells of the nonvirulent strain by a substance that could withstand both the killing and the dissolving of the cells originally containing it. The name "transforming principle" was given to this substance although its chemical nature was not yet known.

This work of Griffith and those that followed it up in the late 1920's and early 1930's was of interest to other biologists but its significance was not appreciated at the time. It wasn't until around 1941 that Oswald T. Avery (1877-1955), Maclyn McCarty (1911-   ), and Colin M. MacLeod (1909-1972) of the Rockefeller Institute followed up Griffith's work and attempted to isolate the transforming principle and determine its chemical nature.

They grew large numbers of the virulent pneumococci which were later heat-killed, washed, centrifuged and broken apart. The resulting products were then extracted, salted, acidified, treated with chloroform, precipitated, redissolved, dried, frozen, or treated in other different ways. From these various operations many cell extracts were produced. The next task was to separate the individual chemical components of these extracts, purify them, and determine which one might be the actual transforming principle. Each suspected chemical had to be tested on non-virulent cells in order to ascertain its ability to change them to the virulent form. Avery and his co-workers gradually narrowed the field to DNA.

George W. Beadle (1904-   ) and Edward L. Tatum (1909-1975) carried on research on the chemistry of pigment formation in the eye of *Drosophila*. They believed that pigment synthesis proceeded by a series of separate steps, each under control of a separate enzyme produced through the influence of a specific gene. In order to prove this, they thought that a new approach was needed. Instead of starting with a mutation and tracing its chemistry back step by step, they decided to search for gene mutations that affected the known chemistry of a process. This simple change of direction was most fortunate since it led to the establishment of biochemical genetics and the Nobel Prize.

By 1940 Beadle and Tatum decided that an organism other than *Drosophila* was needed in order to study gene action. They chose the red bread mold *Neurospora*. This organism could be grown on a specific culture medium consisting of water, salts, sucrose and biotin and from these molecules it synthesized all its needed more complex compounds. They then proceeded on the assumption that X-ray treatment would induce mutations in genes concerned with the control of known specific chemical reactions. Beadle and Tatum reasoned that *Neurospora* had to be able to carry out a certain chemical reaction in order to survive on the regular culture medium while a mutant unable to do this would die on the same medium. Such a mutant would be able to be maintained, though, by growing it on a medium to which had been added the essential product of the genetically blocked reaction.

In order to test this hypothesis, X-rayed *Neurospora* cultures were established on the regular culture medium to which all known vitamins and amino acids and several alternative carbon sources had been added. Subsequently these *Neurospora* were then transferred to a special medium that required the organism to carry on all of the essential syntheses of which it was capable. Any loss of the ability to synthesize an essential substance present in the regular medium and absent in the special one was indicated by a strain growing on the first and failing to grow on the second. These strains were then tested in a systematic manner to determine what substance they were unable to synthesize. For example, the 299th spore of *Neurospora* that Beadle and Tatum had irradiated produced a mutant strain that needed vitamin B_6 to survive while the 1,090th spore required vitamin B_1.

From these observations emerged the hypothesis that the gene produces its effect through enzymes. Normal *Neurospora* contained all of the enzymes needed to produce its required compounds and exposure to X-rays caused some genes to mutate knocking out some of the enzymes.

In his Nobel Prize lecture, Beadle admitted that he and Tatum had merely rediscovered something that Garrod had indicated many years before. He pointed out that Garrod clearly had in mind the concept of a gene-enzyme-chemical reaction system. Despite the clear-cut nature of Garrod's work, it had little influence on the thinking of his contemporaries and it wasn't until fifty years later with the work of Beadle and Tatum that his hypothesis received verification.

A number of years went by after the demonstration by Avery and his group that DNA could transform harmless pneumococci into virulent organisms but geneticists still refused to get excited over DNA. They were well satisfied with the gene theory built on a firm foundation of experiments and the hypothetical gene seemed quite adequate to explain all phenomena thus far observed. They held the basic assumption that proteins played a key role in heredity.

Strong evidence that DNA was the hereditary material came from the studies of Alfred D. Hershey (1908-   ) and Martha Chase 1927-   ). They designed an experiment to determine whether an infecting bacteriophage injected into *E. coli* bacteria only DNA, only protein, or a combination of both. Phages were cultured on bacteria that were grown on a medium containing radioactive isotopes of phosphorous and sulfur. The phages incorporated this sulfur into their protein and the phosphorous into their DNA. Nonradioactive bacteria were then infected with these radioactive phages. After the phages became attached to the walls of the bacteria and injected their hereditary material, the bacteria were agitated in a blender in order to detach the phage parts connected to the bacterial walls.

Analysis of these parts revealed that they contained radioactive sulfur but not phosphorous, indicating that only the protein coat remained outside the bacterial cell. Analysis of the bacteria showed only the presence of the radioactive phosphorous indicating that only DNA had been injected into the bacteria by the phages. Thus the experiment demonstrated that DNA alone was sufficient to transmit to the bacteria the genetic information necessary for the production of new phages. These results, reported in 1952, supported the earlier conclusions based on Griffith's transformation experiments that nucleic acids constituted the genetic material.

By the early 1950's it had become strikingly clear that heredity was tied up in some way with DNA. The evidence from bacterial and viral genetics (Griffith, Avery, and Hershey) was extremely convincing. Also, by this time, a great deal was known about the chemical composition of the nucleic acids. This information had been gained from chemical analysis carried out in the early 1900's involving the breakdown of DNA into its smaller component parts. With all evidence pointing to DNA as the genetic material, the 1950's was characterized by work being carried on at a furious pace in an effort to understand DNA.

James D. Watson (1928-   ) was influenced by Muller to study the genetics of viruses. He later was influenced by the physicist-turned-biologist Max Delbruck (1906-   ) to

James D. Watson

think of biology in terms of mathematics and physics. At this time the biggest problem around for a biologist was the nature of the hereditary material. Because the Cavendish Laboratory in England was involved in X-ray crystallography, Watson went there and fell in with the British physicist-turned-biologist Francis H. C. Crick (1916-   ). Both men had a mounting enthusiasm for hunting down the elusive DNA structure. They approached the problem in a way conventional in other sciences but novel for biologists, namely the building of a model of the structure being studied.

The most important clues that they had for guidance were the X-ray pictures taken by Maurice Wilkins (1916-   ) and Rosalind Franklin (1920-1958). When DNA is extracted from cells, the purified material appears jelly-like. Wilkins pulled a delicate fiber of DNA from one such gel prepared from cells of the bacterium E. coli. When he examined this fiber microscopically, he detected some indication of a crystalline structure. This prompted him to attempt analysis by X-ray diffraction. The pictures obtained by this process suggested that the DNA molecule had a regular structure in the shape of a spiral. The pattern was the same no matter what the species of DNA.

Watson and Crick devoted themselves to building models to fit the data established by the X-ray diffraction work of Wilkins. Rods, clamps and bits of sheet metal cut to represent DNA components were assembled like a jigsaw puzzle. Consideration was given to every turn, angle, size and shape indicated by the X-ray diffraction data.

Linus Pauling (1901-   ) also was working on a structure for DNA at this time. Watson was friendly with Pauling's son, who had received a manuscript on DNA from his father. Watson happened to see this and spotted an error that Pauling had made. He took hasty advantage of this in his own search for the structure of DNA.

In eighteen months Watson and Crick found the double helix structure of the DNA molecule. This molecular structure was demonstrated to be correct by Wilkins who had refined his X-ray diffraction process to such precision that it could pinpoint the various atoms within the molecule. Three definitive papers published together in the British scientific journal *Nature* in 1953 sent waves of excitement through the entire world community of biological scientists. These were "The Molecular Structure of DNA" by Wilkins et. al., "A Structure of DNA" by Watson and Crick, and "Genetical Implications of the Structure of DNA" by Watson and Crick. No fundamental revision of the Watson Crick model has been found necessary since these papers first appeared.

Genetics in the Soviet Union

The situation in biological science that prevailed in the Soviet Union from 1937 to 1964 represents a most unusual chapter in the history of modern science. During that period, total control over both research in biology and practical agriculture was entrusted to a charlatan, Trofim D. Lysenko (1898-1976). How this could have happened in a country that rivaled the United States in developing a nuclear potential and that was in the forefront of space exploration is still only an area of keen speculation.

Lysenko was trained as an agronomist and did not receive a rigorous biological education. He started his career as a junior plant breeding specialist in a rural experimental agricultral station. At a time that food production on Stalin's new collective farms was disastrously low, Lysenko claimed to have developed a method of treatment for winter wheat seeds that would facilitate their being planted in the spring. A series of similar claims followed, resulting in Lysenko acquiring the support of Stalin. With this, he soon became president of the Lenin Academy of Agricultural Sciences and editor of its scientific journal. His supporters were appointed without election to the Academy of Sciences and they collected an assortment of posts, titles and medals. Those who opposed Lysenko were dismissed from their positions, exiled, imprisoned or executed. Medical genetics, dealing with hereditary diseases, and eugenics were singled out for special abuse by the Lysenkoists. The Medico-Genetical Institute was closed and studies of human genetics were suspended.

Lysenkoism rejected the role of chromosomes and DNA in heredity and accepted the inheritance of acquired characteristics. Mendelism was equated with facism and ridicule and caricature were used to discredit Mendel's contributions. That Mendel was a monk was used as justification for considering his principles invalid. Lysenko maintained that mathematical analysis could not be applied to biology. Uncontrolled experimentation was the basis for fantastic claims such as the transformation of small white fowl into large black ones through blood transfusions or the transformation of beets into cabbage. Falsified data and doctored photographs were employed to support these claims.

In the 1960's, molecular biology and genetics assumed dominant positions in biology, which by then was receiving universal attention as a result of the cracking of the genetic code and the discovery of protein synthesis. Soviet biology, on the other hand, now contrasted sharply with biology in the west which was the scene of these major breakthroughs. No genetics articles were being published in the Soviet Union and the control of biological literature was so tight that it was difficult for articles critical of Lysenko to appear in the Soviet press.

The point was finally reached when the Soviet Union could no longer ignore the progress that was being made in biology outside of the Soviet Union. In 1962 a commission was created that gave legal scope to a return to real genetics. In 1964 Khrushchev was censured and forced to resign, one of the reasons being his unconditional support of Lysenko. One year later, Lysenko was dismissed from his position as director of the Institute of Genetics and publicly disgraced.

The stage was now set for the re-establishment of genetics in the Soviet Union after a hiatus of thirty years. It was necessary to suspend the teaching of biology for one year in order to formulate new curricula, write new textbooks and retrain teachers so that biology education would reflect the advances that were made in the west during these thirty years. Lysenko's appointments had to be removed from their leadership positions in scientific institutes and laboratories and from the editorial boards of biological journals and new competent replacements found. Finally, new research establishments had to be organized in the long neglected disciplines of genetics, biochemistry and molecular biology.

Synthesizing Nucleic Acids

The laboratory synthesis of nucleic acids began in 1955 when Severo Ochoa (1905-   ) produced RNA in the laboratory by using enzymes that stimulate the replication process.

Arthur Kornberg (1918-   ), who was Ochoa's student, did the same for DNA. He is a physician-turned-biologist who believes that any chemical reaction carried on in a cell can be duplicated in a test tube if the correct materials and the proper conditions are provided. He was attempting the biosynthesis of the four nucleotides found in DNA when the Watson Crick model came on the scene. With this model as a guide, Kornberg became involved in determining the chemical mechanism by which DNA is built up in the cell. The reaction system that he successfully developed for the synthesis of DNA required equal amounts of the four nucleotides, triphosphate in the form of ATP, deoxyribose sugar, DNA-synthesizing enzyme from purified *E. coli* bacteria and a small amount of DNA.

In 1961 Marshall Nirenberg (1927-   ) and J. Heinrich Matthaei reported that they had found a way to decipher the coded information in DNA. Nirenberg, in a decisive experiment, had shown that a sequence of nucleotides containing only the base uracil resulted in the formation of a protein containing exclusively the amino acid phenylalanine. With other evidence indicating that a sequence of three bases was responsible for the placement of a single amino acid in a protein, this experiment showed that three uracils in the language of nucleic acids was equivalent to phenylalanine in the language of proteins.

In 1976 H. Gobind Khorana (1922-   ) reported the synthesis of a functional gene consisting of 199 pairs of nucleotides. His approach was to take the four basic nucleotides, each of which is commercially available in purified form, and assemble them into a double stranded DNA molecule whose sequence of units was identical to that known for a natural gene. The first fifty-two nucleotide pairs served as the promotor, starting the natural formation of the gene's product and also determining how often or when the process was to be repeated. The next 126 pairs specified that gene's product, in this case tyrosine transfer RNA. The last twenty-one functioned as the terminator, ordering the halt to the formation of a new molecule.

Recombinant DNA and Genetic Engineering

Recent biological research involving the recombinant DNA technique has provoked public concern as to the potential human risks associated with genetic engineering. The concept of genetic engineering is to insert new genes into the hereditary material of a living cell, changing the cell's traits and those of its progeny. The controversy on recombinant DNA research centers on the danger of creating new strains of harmful and infectious DNA elements whose biological properties could not be completely predicted. It is possible that these hybrid DNA molecules created by the new methods could lead to the transfer of novel infections or even cancer viruses to man.

The research involves the use of enzymes to break up or stitch together pieces of DNA which can then be inserted into different cells such as bacteria. These "restriction enzymes" have the property of recognizing specific sequences of DNA nucleotides and cutting the double-stranded molecules in such ways so that other strips of DNA can be joined to them. In many types of bacteria there are auxillary rings of DNA called plasmids in addition to the DNA that is the cell's main genetic apparatus. It is into these plasmids that scientists have been inserting strips of DNA from animal cells. Those bacteria with the animal genes in them can be regarded as man-made organisms. Sometimes a special type of virus called *Lambda* that invades bacteria and incorporates its own genes among the bacterial genes is used instead of plasmids in order to introduce foreign genes. The potential hazard of these experiments is further increased because the bacteria used most often are a special laboratory strain of the intestinal bacterium *E. coli.*

Scientists see many potential benefits to genetic engineering work. They seek to build up large supplies of particular genes they want to study. They also hope to use bacteria to grow substances like human growth hormone for individuals who lack the ability to make their own. Other hopes are to simplify the manufacture of antibiotics, open up another way to obtain insulin for diabetics, and improve the capacity of soil organisms to fix atmospheric nitrogen.

In 1976 an international conference of biologists decided to formulate tighter professional standards governing genetic engineering research. These guidelines, rare in the history of science, represent an attempt to balance the potential hazards and benefits of the new techniques for manipulating genes.

An equally novel development without precedent in the history of science has been the proposal by American scientists to establish a "science court" to hear cases in which the experts disagree on scientific facts. In such a court the adversaries would be able to direct their best arguments at each other and at a panel of sophisticated scientific judges in order to resolve a scientific controversy.

DEVELOPING A THEORY OF EVOLUTION

Setting the Stage for Darwin
Charles Darwin
The Social and Intellectual Climate
Supporting Darwin: Wallace, Huxley and Gray
Darwin's Later Work
Darwinism Today

Setting the Stage for Darwin

Aristotle (384-322 B.C.) arranged living things in a "ladder of nature" and was moving towards a concept of evolution. Erasmus Darwin (1731-1802), an English philosopher and the grandfather of Charles, believed that species descended from common ancestors. He suggested a struggle for existence and speculated about variations in animals and reproduction of the strongest. Charles called his grandfather's ideas "more speculative than scientific."

The book written by George Cuvier (1769-1832) in 1812 about fossils he studied in Paris marks the beginning of paleontology. Cuvier believed that the world had gone through a series of catastrophes—earthquakes, floods, and volcanic eruptions. At these times all living things were destroyed but after each catastrophe new organisms were created. According to him, this would account for the differences in species observed in the lower and upper strata of rocks.

The English geologist Charles Lyell (1797-1875) in 1830 wrote *Principles of Geology—Being an Attempt to Explain the Former Change of the Earth's Surface by Reference to Causes Now in Operation*. He concluded that the age of the earth was greater than anyone had supposed and that the same forces have always, and in approximately the same degree as in the present time, been operating on the earth's surface. He thought that it was up to the believers in the catastrophe theory to prove their view correct. He was also convinced that the phenomena of past ages should be explained on the basis of that which is known about the phenomena of the present. Lyell introduced the principle of starting from what is known and proceeding gradually towards the unknown and is considered to be the founder of modern geology and a pioneer in the development of the descent theory.

Jean Lamarck (1744-1829), a self-taught biologist, first became a professor at the age of fifty-five in a subject in which he never received any formal instruction. It seemed to him improbable that species were unchangeable or that any group became extinct through abrupt catastrophes. Instead, Lamarck believed that species were subject to progressive development from simple to complex. The mechanism that he proposed was the "Law of Use and Disuse." According to this, change of environment led to special demands on certain organs. Those being specifically exercised became specially developed, with this development being transmitted to the offspring.

Cuvier's theory of catatrophism and successive new creations was popularized at the expense of the alternative theory of gradual change supported by Lamarck. Cuvier commanded wide respect in the scientific community and was held in esteem by the French government. As a consequence, he was influential in destroying the effect of Lamarck's theory on contemporary thinking. Lamarck was no match for this younger stronger critic who had prestige, wealth and eloquence. Thus his published works, the most important of which was *Philosophical Zoology*, attracted little attention in his own day or in the immediately succeeding period and were regarded as being fantastically speculative. It was only after Darwin's work, when biologist's were searching for precursors to his theory, that they saw in Lamarck's theories the basis for the correct interpretation of evolution.

Charles Darwin

Charles Darwin (1809-1882) was studying theology in England in 1831 when he was offered the unsalaried post of naturalist on the ship Beagle which was to go around the world for geographical purposes. A last minute sailing gift which Darwin took with him was the newly published *Principles of Geology* by Lyell. The young Darwin was a firm believer in the Christian faith and as such accepted the dogma of divine creation. During the Beagle's voyage, this came into conflict with his observations.

Darwin examined a fossil bone from the head of a giant anteater that he had unearthed in South America and noted that it was different from, though related to the modern forms. Mice that he collected on the east side of the Andes differed from, and yet resembled, those captured on the west side.

He observed thirteen distinct species of finches on the Galapagos Islands and they were all clearly related. He wrote, "One might almost fancy that from an original paucity of birds in this archipelago, one species had been taken and modified for different ends."

"One is astonished at the amount of creative force, if such an expression may be used, displayed on these small, barren and rocky islands and still more so at its diverse yet analogous action on points so near each other."

"I never dreamed that islands, about fifty or sixty miles apart, and most of them in sight of each other, formed of precisely the same rocks, placed under a quite similar climate, rising to a nearly equal height would have been differently tenanted."*

As the Beagle sailed around the world, Darwin worked on the journal describing the geological and zoological features of the countries he visited that he was writing from notes jotted down in the field notebooks he always carried with him. In 1839, this journal, which was dedicated to Lyell, was published and it received a very warm reception.

After the voyage of the Beagle, the subject of species began to haunt Darwin. He thought that if he collected all the facts that bore in any way on variation in animals and plants under domestication and in nature "some light might be thrown on the whole subject." Working on the principle of collecting all facts without regard to theory,

*Journal of Researches Into the Natural History and Geology of the Countries Visited During the Voyage of H.M.S. "Beagle" Round the World. Charles Darwin. Harper. New York. 1846.

Darwin started to gather a massive amount of material and analyze and study the notes and materials that he had collected during the Beagle's voyage.

In order to demonstrate how domestic cattle, horses, dogs, cats, pigeons, grains, flowers and fruits had developed from earlier different ancestors, Darwin delved into history, ancient literature, art, breeder's records and comparative anatomy and also carried on extensive experiments of his own. He traced the dog back to several wolf-like ancestors, horses to a dun-colored wild ancestor, the peach to a modified almond, and wheat to a small grained plant.

He gave his closest attention to the pigeon. Of all the domesticated animals, its ancestry could be traced most fully. He found that the one hundred and fifty modern species went back to the bluish rock pigeon that once had nested among the rocks on Asiatic shores. The modern forms differed so radically both from their ancestors and from each other that Darwin had difficulty persuading even himself that all of them had arisen since man first domesticated the rock pigeon five thousand years earlier. To gather the necessary evidence he set up an aviary of his own, joined two societies of pigeon fanciers and subscribed to breeders' journals and catalogues.

Darwin was convinced that selection was the key to man's success in producing useful varieties of plants and animals. But he could not figure out how selection would work in nature in the absence of the breeder. This remained a mystery to him until 1838 when he happened to read "for amusement" the already famous book of Thomas Malthus (1766-1834) called *Essays on the Principle of Population*. Malthus warned that if the human population continued to increase unchecked, it would ultimately outrun its food supply. He believed that this didn't happen, though, because war, disease and famine were holding down the population. Malthus emphasized the ruthless, continuous struggle occurring in nature.

Darwin thought that this struggle itself could be the selective agent that he was looking for. He wrote, "It at once struck me that under these circumstances the most favorable variations would tend to be preserved and the unfavorable ones destroyed." He further reasoned that as the best adapted survived and the others became extinct, the survivors would eventually differ from their ancestors and develop into a new species.

Darwin now had a theory that held together logically. However, it seemed so radical to him and so at variance with the prevailing thought that he decided against writing it down then. He would think about it and test it in every possible way. It wasn't until 1842 that he wrote a brief abstract of his theory in pencil that consisted of thirty-five pages.

The following are examples of how Darwin attempted to experimentally illustrate the facts and deductions that comprised his theory. One spring he measured off a two by three foot piece of ground and had it dug up and cleared. Each day he marked every seedling that came up. By the end of the season 357 seedlings had grown but 295 of them were destroyed by insects. Thus 83% of the weeds had perished as a result of the struggle for existence.

The question of geographical distribution presented problems for Darwin. In order to demonstrate how seeds could have gotten to the Galapagos Islands he put different kinds of seeds in bottles of salt water for varying periods of time. He found that sixty-four out of eighty-seven kinds of seeds could germinate after being immersed for twenty-eight days. He tested fruits for floating and found that many did. He fed soaked seeds to fish and then fed these fish to storks. The seeds germinated even after being passed with the storks' wastes. When Darwin watered a ball of earth taken from the leg of a bird, he found that eighty-two plants developed from it. He dangled the feet of a duck in his aquarium when many ova of fresh water molluscs were hatching. Thirty of the extremely small just hatched shells climbed on to the duck's feet and could not be removed. They survived there, out of water, for twenty-four hours. In that time a duck could fly at least six hundred miles and if blown across the sea to an oceanic island could come down on a pond.

In January 1844 Darwin wrote, "I have been now ever since my return engaged in a very presumptuous work and I know no one individual who would not say a very foolish one. I was so struck with the distribution of the Galapagos organisms etc., and with the character of the American fossil mammals etc., that I determined to collect blindly every sort of fact which could bear in any way on what are species. At last gleams of light have come, and I am almost convinced (quite contrary to the opinion I started with) that species are not (it is like confessing a murder) immutable."*

During the summer of 1844, Darwin enlarged the 1842 abstract of his theory from the original thirty-five pages to two hundred and thirty pages. It was divided into two parts; I *On the Variation of Organic Beings under Domestication and in their Natural State* and II *On the Evidence Favorable and Opposed to the View that Species are Naturally Formed Races Descended from Common Stocks.*

However, species did not preempt all of Darwin's time. He corrected and revised a new edition of his original 1839 Journal which came out in 1845. He also worked for four

*Charles Darwin—Autobiography and Letters. Edited by his son Francis Darwin. D. Appleton and Company. New York. 1893.

and one half years on his three geological books: *Coral Reefs, Volcanic Islands*, and *Geological Observations on South America*.

In 1855 Alfred R. Wallace (1823-1913), who was working in the East Indies, published an article, "On the Law Which Has Regulated the Introduction of New Species," which argued in favor of evolution. Disappointed that his ideas had attracted so little attention, Wallace wrote to Darwin, having heard that he was interested in the subject, soliciting his opinion. Darwin responded encouragingly.

Darwin's friends always feared the eventuality that someone would come out with a theory similar to his before he publicly announced it and they encouraged him to publish as soon as was possible. But Darwin would not increase his pace. By 1857 it was twenty years since he had opened his first notebook on species. Spurred on by this twenty year anniversary as well as by his friends, Darwin began to pull his materials together and write.

On June 18, 1858 the mail brought to Darwin an essay by Wallace, who was living in Malaya at the time. It was called, "On the Tendency of Varieties to Depart Indefinitely from the Original Type." What Darwin read stunned him. There in a few pages was all of his theory. It was as if Wallace had worked from Darwin's 1844 abstract to the extent that terms used by Wallace appeared as the heads of Darwin's chapters. Here again was another of those classic coincidences of science where two men in different parts of the world independently arrived at the same conclusion.

In letters to Lyell Darwin wrote, "Your words have come true with a vengeance—that I should be forestalled. You said this when I explained to you here very briefly my views of 'Natural Selection' depending on the struggle for existence."

"I never saw a more striking coincidence... So all my

originality, whatever it may amount to, will be smashed, though my book, if it will ever have any value will not be deteriorated; as all the labor consists in the application of the theory."

"There is nothing in Wallace's sketch which is not written out much fuller in my sketch, copied out in 1844 and read by Hooker some dozen years ago. About a year ago I sent a short sketch of which I have a copy, of my views to Asa Gray, so that I could most truly say and prove that I take nothing from Wallace. I should be extremely glad now to publish a sketch of my general views in about a dozen pages or so; but I cannot persuade myself that I can do so honorably. Wallace says nothing about publication, and I enclose his letter. But as I had not intended to publish any sketch, can I do so honorably, because Wallace has sent me an outline of his doctrine? I would far rather burn my whole book, than that he or any other man should think that I had behaved in a paltry spirit. Do you think his having sent me this sketch ties my hands?"

"If I could honorably publish I would state that I was induced now to publish a sketch (and I should be very glad to be permitted to say, to follow your advice long ago given) from Wallace having sent me an outline of my general conclusions. We differ only in that I was led to my views from what artificial selection has done for domestic animals. I would send Wallace a copy of my letter to Asa Gray, to show him that I had not stolen his doctrine."

"Forgive me for adding a P.S. to make the case as strong as possible against myself. Wallace might say 'You did not intend publishing an abstract of your views till you received my communication. Is it fair to take advantage of my having freely, though unasked, communicated to you my ideas, and thus prevent me forestalling you?' The advantage which I should take being that I am induced to publish from

Charles Darwin during the period when he was writing *On the Origin of Species by Means of Natural Selection*.

privately knowing that Wallace is in the field. It seems hard on me that I should thus be compelled to lose my priority of many years' standing, but I cannot feel at all sure that this alters the justice of the case. First impressions are generally right, and I at first thought it would be dishonorable in me now to publish."*

These words to Lyell indicate a very confused Darwin, torn between his desire for recognition and his fear of being dishonorable. This situation was finally resolved when Darwin's friends decided that a joint presentation of both Darwin's and Wallace's works was absolutely essential. Thus the following four documents were presented at the July 1, 1858 meeting of the Linnean Society:

1. A letter from Lyell and Hooker explaining how Darwin happened to receive Wallace's paper and describing his first impulse to publish it. They indicated that they had not permitted him to do this and that both men had reached their conclusions independently with neither knowing what the other was doing.

2. Extracts from Darwin's unpublished paper of 1844. Hooker appended a statement that he had read it the year that it was written.

3. The letter written by Darwin to Asa Gray in which he stated his views.

4. Wallace's paper.

Since the Linnean Society meeting had originally been scheduled to hear a paper asserting that species were fixed, the words of Darwin and Wallace came as a surprise. Because no discussion followed the reading, Joseph D. Hooker (1817-1911), a botanist and lifelong intimate friend of Darwin's, thought that those who were opposed to the new

*Charles Darwin—Autobiography and Letters. Edited by his son Francis Darwin. D. Appleton and Company. New York. 1893.

theories were not prepared to present their arguments without proper preparation.

Darwin was very pleased with the joint reading and began to prepare the abstract that all his friends agreed he had to write immediately. Although suffering from headaches and a bad stomach, he managed to complete the manuscript in thirteen months. Darwin had been able to use only a fraction of his stored materials, and was concerned that readers would have to take some of his statements on trust without seeing the substantiating evidence. Finally, on November 24, 1859 *On the Origin of Species by Means of Natural Selection* was published. The 1,250 copies printed for the first edition were sold on the first day and a second edition of 3,000 copies was ordered immediately.

The Social and Intellectual Climate

Liberalism originated in England. Throughout the eighteenth century humane legislation and democratic social development were demanded and gradually achieved without violent upheavals. Free competition and no interference with the individual's liberty of action existed. The new big scale industrial development and world trade created an intelligent middle class which felt well satisfied with the present and hoped for still greater benefits from the future. There was belief in the natural goodness of the human race and in evolution and progress. Darwin's theory was compatible with this political liberalism of the time in that it was viewed as an application of the doctrine of free competition to the natural world.

The British engineer Herbert Spencer (1820-1903) built a system of philosophy based on evolution. He wanted to combine the results of physics, chemistry, biology, psychology and sociology and thought that this could be done through evolution which unified all the sciences. In 1852 Spencer first used the term evolution to describe the general process by which higher forms were produced from lower ones. Darwin hardly used the word but Spencer's use of it caught on rapidly and eventually Darwinism and evolution were considered to have the same meaning. Because Spencer's works were widely read, they were influential in spreading the concept of evolution. The phrase "survival of the fittest" also was coined by Spencer and quickly became popular.

During the decades following Darwin's work, Germany assumed a leading position in biological research. The great technical and economic development that resulted from the founding of the German Empire facilitated research and the work at German universities was well organized and characterized by careful guidance given by the teachers to their pupils' scientific work. Comparative anatomy was especially highly developed there and German biologists involved themselves in finding support for Darwin's theory from comparative anatomy and embryology.

Darwinism was least appreciated in France, probably due to the influence of Cuvier. Also, representatives of experimental research like Bernard were not attracted by the hypothetical and speculative elements in Darwin's theory.

Supporting Darwin: Wallace, Huxley and Gray

Alfred R. Wallace developed into an amateur naturalist through his own reading and studies. Darwin's Journal describing the voyage of the Beagle greatly impressed him and fostered his desire to visit distant places. Wallace pursued his interest in nature by catching insects and selling

Alfred R. Wallace

them to collectors. The money that he received was used to finance a trip to Brazil where he traveled extensively and visited the most remote areas. It was here that Wallace contracted a tropical disease characterized by a recurrent fever that was to plague him for many years after.

After writing *A Narrative of Travels on the Amazon and Rio Negro* Wallace decided to explore the Malay Archipelago. In evenings, on rainy days, or when in the midst of a siege of the fever contracted in Brazil, all times not actively involved in the collection of specimens, Wallace speculated about the origin of species. In 1855 he wrote to Darwin, soliciting his opinion of a paper entitled "On the Law Which Has Regulated the Introduction of New Species." In it, Wallace argued in favor of evolution. Darwin's favorable response greatly encouraged Wallace and the two began an occasional correspondence. In the midst of one bout with the tropical fever Wallace thought of Malthus and his *Essay on Population* which he had once read. It seemed reasonable to Wallace that the checks on increases in human populations referred to by Malthus might also apply to animal populations. He wrote, "It suddenly flashed upon me that this self-acting process would necessarily improve the race, because in every generation the inferior would inevitably be killed off and the superior would remain—that is, the fittest would survive."

Initially Wallace was too weak from the fever to write down his thoughts, but as he recovered he began to write out the ideas that had come to him. In one week he had completed the paper, "On the Tendency of Varieties to Depart Indefinitely from the Original Type," and mailed it to Darwin. Darwin's receipt of this paper precipitated the joint presentation of both his and Wallace's works at the meeting of the Linnean Society on July 1, 1858.

Wallace returned to England in 1862 and he and Darwin kept in regular contact through visits and correspondence. Wallace also maintained close relationships with many prominent scientists, including Huxley, Hooker and Galton. In 1876 he published a book on the geographical distribution of land animals, establishing biogeography as a science.

In 1886 Wallace lectured in Boston on "The Darwinian Theory" and a local newspaper called his presentation "a masterpiece of condensed statement—as clear and simple as compact—a most beautiful specimen of scientific work." Because of his ability to simplify complex ideas, Wallace contributed greatly to the acceptance of evolution in the United States. Eventually he wrote *Darwinism* which he thought would "enable any intelligent reader to obtain a clear conception of Darwin's works."

In 1908, on the fiftieth anniversary of the reading of Darwin's and Wallace's papers, the Linnean Society established the Darwin-Wallace Medal and chose Wallace as its first recipient. In his acceptance speech, Wallace explained why he had always modestly stepped aside in favor of Darwin when it came to receiving recognition for their theories. He indicated that Darwin was determined not to publicize his conception until it could be backed up with substantial proof while in his case the idea came to him in a sudden flash of insight, was thought out, written down and then mailed to Darwin, all in one week. Thus Wallace believed that the credit should be proportional to the time devoted to the problem—twenty years for Darwin to one week for him. Nonetheless, when Wallace died, a medallion honoring him was placed in Westminster Abbey next to one honoring Darwin.

Thomas H. Huxley (1825-1895) was a British biologist who was the grandfather of Julian and Aldous. He lectured

A caricature of Thomas H. Huxley that appeared at the time he was defending Darwin's Theory.

and published in virtually all areas of biology and was also interested in the improvement of elementary and secondary education. Although in his youth he was an upholder of the immutability of species and an opponent of Lamarck's theory, he gradually became converted to Darwinism and became one of its most zealous champions. In fact, he was nicknamed "Darwin's bulldog." At a meeting of the British Association for the Advancement of Science, Samuel Wilberforce, the Bishop of Oxford, after attacking Darwin inquired of Huxley, "Was it through his grandfather or his grandmother that he claimed descent from an ape?" Huxley retorted, ". . . that a man has no reason to be ashamed of having an ape for his grandfather. If there were an ancestor whom I should feel shame in recalling, it would be a man, a man of restless and versatile intellect, who, not content with an equivocal success in his own sphere of activity, plunges into scientific questions with which he has no real acquaintance, only to obscure them by an aimless rhetoric, and distract the attention of his hearers from the real point at issue by eloquent digressions, and skilled appeals to religious prejudice."*

Asa Gray (1810-1888) was a botany professor at Harvard University who supported Darwin in opposition to his own colleague, Louis Agassiz (1807-1873). Gray wrote a series of articles defending Darwin in the *Atlantic Monthly* and *Darwiniana*, a collection of his views in favor of Darwin's theory.

Agassiz was a famous lecturer and popularizer of science who exercised a national influence on American science. He became the principal critic of Darwinism in the United States, using his great wealth of information to point out

---

*Charles Darwin—Autobiography and Letters. Edited by his son Francis Darwin. D. Appleton and Company. New York. 1893.

what he regarded as the fallacies and weaknesses of Darwin's theory. However, because of his unyielding position as he made his way through the ensuing scientific debates, his brilliant scientific reputation was eventually tarnished.

Darwin's Later Work

Darwin believed that the same biological principles applied in the same way to man as well as to all other living things. He therefore collected notes relating to the evolution of man but only for his own satisfaction. He did not intend to publish them because of the controversy he knew they would arouse. However, in order not to conceal his views, Darwin stated in the *Origin* that from its contents "light would be thrown on the origin of man and his history." It was this passing reference to man that was severely twisted and misinterpreted and that evoked the furor that followed the *Origin's* publication.

When both Lyell and Wallace indicated no interest in pursuing the issue of man, Darwin began to look over his voluminous notes and found three classes of facts—structural, embryonic and vestigial—that indicated descent from a common ancestor.

After observing that secondary sexual characteristics possessed by males of most species served to give some males advantages over others, Darwin began a colossal study of sexual selection. In 1871 after three years' work, he completed *The Descent of Man and Selection in Relation to Sex*. It contained six chapters on man and fourteen on sexual selection. With the publication of this work the opposition to Darwin's views flared even higher.

Next Darwin began to pull together his material on emotional behavior and completed *The Expression of Emotions in Man and Animals*. In *Descent* he had relied heavily on evidence that lower animals experienced the same emotions as man and expressed them in much the same way.

In 1875 Darwin finished a book on insectivorous plants in which he showed that the digestive secretions of these plants were very similar to those of animals. Here was still another example of the many interrelationships that he found to exist between the various forms of life. Darwin had proven himself to be an exceptional botanist and during the next few years wrote *The Movements and Habits of Climbing Plants* (1875), *The Effects of Cross and Self Fertilization* (1876), *The Different Forms of Flowers on Plants of the Same Species* (1876), *The Power of Movement in Plants* (1880) and *The Formation of Vegetable Mold Through the Action of Worms* (1881). Many plant physiological investigations of today are based on leads that he uncovered and entire areas, such as studies of hormonally controlled correlations, can be said to have developed directly from his simple experiments and the observations that he made.

When Darwin died in 1882 he received the world's acclaim. He was unique in that he was one of the few scientists who both stated a great principle of nature and also did the massive work required to establish the proof of that principle. Usually in science the filling in of the proof has come long afterward. Subsequently, Darwin's theories fit perfectly into the newly developing science of genetics with modern genetics upholding his work.

Darwinism Today

Some of the controversy generated by the publication of the *Origin* has persisted until recent times. Although most

Americans, including many prominent theologians, accept the validity of the evolutionary theory, there remain a sizable number of persons, including some in positions of influence, who oppose evolution as a false theory and as counter to what they accept as the divinely inspired account in Genesis. It was as late as the 1970's that some states first repealed laws prohibiting the teaching of evolution in public schools. The debate is still being waged in school boards and legislatures around the country and the debate has had an effect on the publishers of biology textbooks for primary and secondary schools. Eager not to lose sales in states where anti-evolution forces affect textbook selection many publishers have omitted or watered down discussion of evolution in their books. Other publishers have included the theory of creation as well as Darwin's theory as an accommodation to the anti-evolution forces.